From Tool to Partner:
The Evolution of
Human-Computer Interaction

Synthesis Lectures on Human-Centered Informatics

Editor
John M. Carroll, *Penn State University*

Human-Centered Informatics (HCI) is the intersection of the cultural, the social, the cognitive, and the aesthetic with computing and information technology. It encompasses a huge range of issues, theories, technologies, designs, tools, environments, and human experiences in knowledge work, recreation and leisure activity, teaching and learning, and the potpourri of everyday life. The series publishes state-of-the-art syntheses, case studies, and tutorials in key areas. It shares the focus of leading international conferences in HCI.

Dedicated to the students who will take our interaction to the next level

From Tool to Partner: The Evolution of Human-Computer
Interaction Jonathan Grudin

ISBN: 978-3-031-01090-3   print
ISBN: 978-3-031-02218-0   ebook

DOI 10.1007/978-3-031-02218-0

A Publication in the Morgan and Claypool Publishers series
SYNTHESIS LECTURES ON HUMAN-CENTERED INFORMATICS, #35
Series Editor: John M. Carroll, Penn State University

Series ISSN: 1946-7680 Print   1946-7699 Electronic

From Tool to Partner:
The Evolution of
Human-Computer Interaction

Jonathan Grudin
Microsoft Research

SYNTHESIS LECTURES ON HUMAN-CENTERED INFORMATICS #35

ABSTRACT

This is the first comprehensive history of human-computer interaction (HCI). Whether you are a user-experience professional or an academic researcher, whether you identify with computer science, human factors, information systems, information science, design, or communication, you can discover how your experiences fit into the expanding field of HCI. You can determine where to look for relevant information in other fields—and where you won't find it.

This book describes the different fields that have participated in improving our digital tools. It is organized chronologically, describing major developments across fields in each period. Computer use has changed radically, but many underlying forces are constant. Technology has changed rapidly, human nature very little. An irresistible force meets an immovable object. The exponential rate of technological change gives us little time to react before technology moves on. Patterns and trajectories described in this book provide your best chance to anticipate what could come next.

We have reached a turning point. Tools that we built for ourselves to use are increasingly influencing how we use them, in ways that are planned and sometimes unplanned. The book ends with issues worthy of consideration as we explore the new world that we and our digital partners are shaping.

KEYWORDS

human-computer interaction, human factors, information systems, information science, office automation, artificial intelligence, history, symbiosis, publication culture, Moore's law, hardware generations, graphical user interface

Contents

Preface

We have reason to celebrate. Half a century ago, inspirational accounts described how computers might someday enable people to accomplish amazing things. Computers at that time were huge, expensive, and could do very little. It took fifty years, but we did it! The early visions have been realized. There is always more to be done, and the next fifty years could be more exciting than the last, but we should recognize what we have collectively accomplished, how we did it, and what this can tell us about the next steps. This book describes what we accomplished, how we did it, and who "we" are. Knowing this should help us take our next steps.

As computers became available, people with different goals looked for ways to make these flexible tools useful: engineers and scientists, businesses and government agencies, hobbyists, librarians and information professionals, and eventually the general public. Different research and development communities formed, in human factors, management information systems, computer science, and library and information science. Most people who are engaged in understanding and improving human-computer interaction associate with one of these fields or with more recent arrivals such as design and communication studies, or they work outside these fields yet draw on findings and methods from them.

It requires an effort to grasp the relationships among these disciplines and discover what could be useful for you. Even the origins and span of your own field may be unclear. To understand what your field is, you must discover what it is not. A good strategy is to see how closely related fields differ. This book was written to help with that. Understanding the past can be the best preparation for a future, despite the surprises that inevitably come.

In the early 1980s, I attended computer-human interaction (CHI) and human factors meetings. I was introduced to the management information systems literature. I wondered why these disciplines did not interact more. Over time, other questions arose: Why didn't the relevant National Science Foundation and the U.S. Department of Defense Advanced Research Projects Agency (ARPA) program managers attend CHI? Why was some of our field's most-acclaimed research ignored by practitioners?

I found answers to these and other questions. By examining similarities and differences in human-computer interaction (HCI) as practiced in computer science, human factors, information systems, and information science, we gain a deeper appreciation of the activity around us. You may find where you could benefit from other fields—and also see where exploration is less promising. Readers in fields that more recently embraced HCI, such as design and communication, can position themselves.

In revising and extending material that I previously published in various handbook chapters, the *IEEE Annals of the History of Computing*, and other places, I realized that we are developing a new relationship with digital devices. It is not a uniform relationship—it appears here and there—but the trajectory is clear even if the benefits and costs are not. The tools that we shaped take the initiative more often than they did in the past, creating new opportunities and new challenges for the designers, developers, and researchers who shift their perspective to explore the possibilities.

Acknowledgments

This book was assembled with contributions from hundreds of people who shared recollections, information, and additional sources. Without the encouragement of Ron Baecker and Phil Barnard, I would not have persisted. Steve Poltrock and I examined much of this ground in courses that we taught. John King and Clayton Lewis were fonts of information and ideas.

My wife Gayna Williams has been a constant partner in exploring and interpreting technology use. She shaped the chapters on recent activity. Seeing my daughters Eleanor and Isobel encounter technology has brought into sharper focus the nature and evolution of our interaction with machines.

Dick Pew generously provided the opportunity to write a history handbook chapter. Don Norman and Brian Shackel patiently responded to questions over the years. Gerhard Fischer, Gary Olson, Mark Tarlton, Steve Poltrock, Glenn Kowack, and Ron Baecker provided insightful comments on drafts and improved the prose immeasurably. Producing a book is a group effort: Diane Cerra's patient guidance, Deb Gabriel's exhaustive editing, Ted Laux's professional indexing, and Susie Batford's wonderful cover illustration, undertaken to improve your experience, had that effect on mine.

I owe special thanks to Don Chaffin and Judy Olson for pointing out the relevance of Lillian Gilbreth and Grace Hopper; no doubt I overlooked people whose contributions deserved inclusion, but I am glad these two made it.

Of the fields covered, library and information science, with roots in the humanities, is home to the most-professional historians. Their welcome to an amateur was wonderful. Exchanges with William Aspray, author of elegant histories of computer science, and Colin Burke, one of the most dedicated scholars I know, washed away any weariness from hours of tracking down information. I hope tomorrow's chroniclers—perhaps you—will draw on this in writing future histories.

CHAPTER 1

Preamble:
History in a Time of Rapid Change

It was 1963. Few computers existed. Joseph Carl Robnett Licklider, an MIT psychologist and engineer, was appointed director at the U.S. Department of Defense Advanced Research Projects Agency (ARPA). He wrote "Memorandum for Members and Affiliates of the Intergalactic Computer Network." The ARPANET, predecessor to the internet, soon followed. It was not Licklider's first sketch of the future. His 1960 article on human-computer "symbiosis" envisioned three stages in the relationship of humans and the mammoth new devices that had far less capability than a graphing calculator does today:

1. Human-computer interaction. Machines are tools, extensions of the arm and eye. Licklider outlined the requirements to make progress: better input and output techniques, better interaction languages. In a 1965 book, *Libraries of the Future*, he provided a more detailed blueprint.

2. Human-computer symbiosis. "It will involve very close coupling between the human and the electronic members of the partnership," which "will think as no human brain has ever thought and process data in a way not approached by the information-handling machines we know today."

3. Ultra-intelligent machines. Assured by leading researchers that machines of greater than human intelligence were coming, Licklider "conceded dominance in the distant future of cerebration to machines alone."

We spent decades in the first stage, in which interaction consisted of a human acting and a computer reacting. We loaded a deck of cards, typed a command, selected a menu item, or clicked on an icon; the computer responded and then awaited our next action.

Reaching the second stage took longer than expected. But without fanfare, the partnership arrived. We've met Licklider's stage 1 goals. It is not an equal partnership. Software does not have a mind, but it can perform tasks that we can't and embody understandings that are based on insights of its human designers and augmented by contextual information that it acquired in use. Now, my devices act on my behalf. Not always and not dramatically, but even as I sleep, they receive, filter, and prioritize information for me. Often I take the initiative; other times my device does, or it offers to. Computer capabilities complement those of a human partner. Digital agents have narrow scopes of action, so their human partners must assess the broader context. Digital partners often have impressive computational skills and clumsy social skills. But these partners are not an alien species; at their best, they represent a distillation of what we have learned so far about technology and ourselves.

This transition profoundly alters the perspectives that will best serve researchers and developers, in ways that we have not yet assimilated. Understanding the paths in designing and understanding interaction that brought us to this point will enable us to identify effective new approaches and to identify practices that are outmoded, as well.

1.1   WHY STUDY THE HISTORY OF HUMAN-COMPUTER INTERACTION?

> *"What is a typewriter?" my six-year-old daughter asked me.*
> *I hesitated. "Well, it's like a computer," I began.*

A widely read 1983 paper advised the designers of a word processor to use an analogy that was familiar to everyone: a typewriter.[1] Already by 1990, a Danish student challenged this reading assignment, arguing "the typewriter is a species on its last legs." My daughters are now teenagers. For them and their friends, typewriters are extinct, as are other species that participated in interaction through the computing era: 80-column punch cards, paper and magnetic tape, line editors, 1920-character monochrome displays, 1-megabye diskettes, CDs, and even DVDs. Are the interaction issues posed by those devices relevant today? No.

In contrast, aspects of the *human* side of human-computer interaction (HCI) change slowly if at all. Much of what was learned about our perceptual, cognitive, social, and emotional processes when we interacted with older technologies applies to emerging technologies. Our reasons for retrieving and organizing information persist, even when the specific technologies that we use change.

This book focuses on interactions with extinct as well as living technologies. There are several reasons for understanding the field's history. Paradoxically, the rapid pace of technology change could strengthen some of them.

1. Several disciplines are engaged in HCI research and application, but few people are exposed to more than one. By seeing how each evolved, we can identify potential benefits of expanding our focus and obstacles to doing so.

2. Celebrating the accomplishments of past visionaries and innovators is part of building a community and inspiring future contributors, even when some past achievements require an effort to appreciate today.

---

[1]   Carroll and Mack 1984.

3. Some visions and prototypes were quickly converted to widespread application, others took decades to influence use, and a few remain unrealized. By understanding the reasons for different outcomes, we can assess today's visions more realistically.

4. Crystal balls are notoriously unreliable, but anyone planning or managing a career in a rapidly changing field must consider the future. Our best chance to anticipate change is to find trajectories that extend from the past through the present. The future will not resemble the present, so it is worth trying to prepare for it.

This account does not emphasize engineering "firsts." It focuses on technologies and practices as they became widely used, as reflected in the spread of systems and applications. This was often paralleled by the formation of new research fields, changes in existing disciplines, and the creation and evolution of professional associations and publications. More a social history than a conceptual history, this survey identifies trends that you might download into your crystal ball. For those interested in deeper exploration of conceptual history, I have gathered some resources in a short appendix.[2]

Not all methodological and theoretical contributions are covered. Some work that strongly influenced me is not covered; nor is my most-cited work. A historical account is a perspective. It emphasizes some things and de-emphasizes or omits others. A history can be wrong in details, but it is never right in any final sense. Your questions and interests determine how useful a given perspective is to you.

A model for intellectual histories of HCI was established by Ron Baecker in the 1987 and 1995 editions of *Readings in Human-Computer Interaction* and was followed by that of Richard Pew (2003) in Jacko and Sears's *Human-Computer Interaction Handbook*. Brian Shackel (1997) summarized European contributions. Perlman et al. (1995) is a compendium of early HCI papers that first appeared in the human factors literature. HCI research in management information systems is covered by Banker and Kaufmann (2004) and Zhang et al. (2009). Rayward (1983; 1998) and Burke (1994; 2007) review the pre-digital history of information science. The incandescent surge of activity in the 1970s and 1980s has been addressed in a wave of popular books, many of them listed in Appendix B.

My contributions to HCI history began with a short 2004 encyclopedia entry and a 2005 article in *IEEE Annals of the History of Computing*. I expanded these into a series of handbook chapters and further explored historical issues in essays that I wrote and invited others to write for a column published in *ACM Interactions* from 2006 to 2013.[3]

Neither I nor most of the authors mentioned above are trained historians. Many of us lived through much of the computing era as participants and witnesses, yielding rich insights and ques-

---

[2] Examples of relevant engineering and conceptual history essays are Myers (1998) on graphical user interfaces and Blackwell (2006) on the use of metaphor in design.

[3] See http://www.jonathangrudin.com/wp-content/uploads/2016/12/Timelines.pdf.

tionable objectivity. This account draws on extensive literature and hundreds of formal interviews and informal discussions, but everyone who participated in past events has biases. Personal experiences can enliven an account by conveying human consequences of changes that otherwise appear abstract or distant and can indicate possible sources of bias. Some readers enjoy anecdotes; others find them irritating. I try to satisfy both groups by restricting personal examples to an appendix, akin to "deleted scenes" on a DVD.

I include links to freely accessed digital reproductions of some early works. When a good Wikipedia article or YouTube video will be found with obvious search engine query terms, I do not interrupt with a full reference; for example, "Three Mile Island nuclear power disaster," "1984 Macintosh Super Bowl video," or "Mundaneum." Similarly, when a sentence identifies the author and year of a contribution, I do not add a citation, but the work appears in the references.

1.2   DEFINITIONS: HCI, CHI, HF&E, IT, IS, LIS

HCI is often used narrowly to refer to work in one discipline. I define it broadly to cover major threads of research and development in four disciplines: human factors, information systems, computer science, and library and information science. Later I discuss how differences in the use of simple terms make it difficult to explore the literature. Here I explain my use of key disciplinary labels. Computer-human interaction (CHI) is given a narrower focus than HCI; CHI is associated primarily with computer science, the Association for Computing Machinery Special Interest Group (ACM SIGCHI), and the latter's annual CHI conference. I use human factors and ergonomics interchangeably and refer to the discipline as HF&E. (Some writers define ergonomics more narrowly around hardware.) The Human Factors Society (HFS) became the Human Factors and Ergonomics Society (HFES) in 1992. Information systems (IS) refers to the discipline within schools of management or business that has also been labeled data processing (DP) and management information systems (MIS), and sometimes information technology (IT). Organizational information systems specialists are referred to here as IT professionals, or IT pros. The Association for Information Systems Special Interest Group on Human Computer Interaction (AIS SIGHCI) should be distinguished from SIGCHI. Library and information science (LIS) represents an old field with a new digital incarnation that includes important HCI research. With "IS" taken by information systems, I do not abbreviate information science, a discipline that often goes by simply "information." "Information schools" proliferated in recent years, sometimes replacing "library schools." A Glossary of these and other acronyms is provided.

1.3   SHIFTING CONTEXT: MOORE'S LAW AND THE PASSAGE OF TIME

A challenge in interpreting past events is keeping in mind the radical change in what a typical computer was from one decade to the next. Conceptual development can be detached from hardware to some extent, but the evolving course of research and development cannot. We are familiar with Moore's law, but we do not reason well about supralinear or exponential growth. People often failed to anticipate how rapidly change would come, and then when it came, they did not credit the role played by the underlying technology. Looking back, we often fail to recall the magnitude of that change.

Moore's law specifies the number of transistors on an integrated circuit; this book considers the broader range of phenomena that exhibit exponential growth. Narrowly defined, Moore's law may be revoked, but the health of the technology industry is tied to ongoing hardware innovation and efficiency gains. Don't underestimate human ingenuity when so much is at stake. Future advances could come through novel materials, three-dimensional architectures, optical computing, more effective parallelism, increased software efficiency, or something unexpected. They will come.

With the arrival of small and specialized devices, the definition of "computer" blurred. I use "computer" and "digital technology" interchangeably. Much of the historical literature forgets to adjust prices to account for inflation. One dollar when the first commercial computers appeared is equivalent to ten dollars today. I converted prices, costs, and funding to U.S. dollars as of 2016. In a few excerpts, I translated archaic 20th-century English into modern 21st-century English by replacing "man" with "human."

As computer use expanded, it attracted the attention of new fields. Think of HCI as streams that started high on several peaks and gathered strength as they flowed to a world-encircling sea. A few engineers and information technologists looked at a lively stream and imagined where it might go. Human factors and management information systems tributaries started. Most streams were nourished by different schools of psychology. Cognitive and social psychologists who saw how to guide an increasingly powerful torrent merged with computer scientists interested in graphics and software engineering. It was not over. Declining digital storage and processing costs fed the library and information science stream and brought a surge of design activity. As millions of people became networked, the discipline of communication expanded into computer-mediated communication and social media. Other fields contributed rivulets. HCI became ubiquitous.

The following chapters fill out this account, organized into eight historical periods and two that reflect on patterns that emerged and identify areas to attend to as we move forward.

CHAPTER 2

Human-Tool Interaction and Information Processing at the Dawn of Computing

In the century prior to arrival of the first digital computers, new technologies gave rise to two fields of research that later contributed to HCI. One focused on making the human use of tools more efficient, the other focused in ways to represent and distribute information more effectively.

2.1 LILLIAN GILBRETH AND THE ORIGINS OF HUMAN FACTORS

Frederick Taylor employed technologies and methods developed in the late 19th century—photography, moving pictures, and statistical analysis—to improve work practices by reducing performance time. Time-and-motion studies were applied to assembly-line manufacturing and other manual tasks. Despite the uneasiness with "Taylorism" reflected in Charlie Chaplin's popular satire *Modern Times*, scientists and engineers continued applying this approach to boost efficiency and productivity.

Lillian Gilbreth and her husband Frank were the first engineers to add psychology to Taylor's "scientific management." Lillian Gilbreth's Ph.D. was the first degree awarded in industrial psychology. She studied and designed for efficiency and worker experience as a whole; some consider her the founder of modern human factors. She advised five U.S. presidents and was the first woman inducted into the National Academy of Engineering.

World War I and World War II gave rise to efforts to match people to jobs, to train them, and to design equipment that was more easily mastered. Engineering psychology was born in World War II after simple flaws in the design of aircraft controls and escape hatches led to plane losses and thousands of casualties.[4] Among the legacies of World War II were respect for the potential of computing, based on code-breaking successes, and an enduring interest in behavioral requirements for design.

During the war, aviation engineers, psychologists, and physicians formed the Aeromedical Engineering Association. After the war, the terms "human engineering," "human factors," and

---

[4] Controls are discussed in Roscoe, (1997); the consequences of ill-designed escape hatches are in Dyson's (1979) chilling account.

"ergonomics" came into use, the latter primarily in Europe. For more on this history, see Roscoe (1997), Meister (1999), and HFES (2010).

Early tool use, whether by assembly-line workers or pilots, was not discretionary. If training was necessary, people were trained. One research goal was to reduce training time, but more important was to increase the speed and reliability of skilled performance.

2.2   ORIGINS OF THE FOCUS ON INFORMATION

Science fiction author H. G. Wells campaigned for decades to improve society by improving information dissemination. In a 1905 non-fiction book he proposed a system based on a new technology: index cards!

> These index cards might conceivably be transparent and so contrived as to give a photographic copy promptly whenever it was needed, and they could have an attachment into which would slip a ticket bearing the name of the locality in which the individual was last reported. A little army of attendants would be at work on this index day and night…
> An incessant stream of information would come, of births, of deaths, of arrivals at inns, of applications to post-offices for letters, of tickets taken for long journeys, of criminal convictions, marriages, applications for public doles and the like. A filter of offices would sort the stream, and all day and all night for ever a swarm of clerks would go to and fro correcting this central register, and photographing copies of its entries for transmission to the subordinate local stations, in response to their inquiries…

Would such a human-powered "Web 2.0" be a tool for social control or public information access? The image evokes both the potential and the challenges of the information era that are taking shape now, a century later.

In the late 19th century, technologies and practices for compressing, distributing, and organizing information bloomed. Index cards, folders, and filing cabinets—models for icons on computer displays much later—were important inventions that influenced the management of information and organizations in the early 20th century.[5] Typewriters and carbon paper facilitated information dissemination, as did the mimeograph machine, patented by Thomas Edison. Hollerith punch cards and electromechanical tabulation, celebrated steps toward computing, were heavily used to process information in industry.

Photography was used to record information as well as behavior. For almost a century, microfilm was the most efficient way to compress, duplicate, and disseminate large amounts of information. Paul Otlet, Vannevar Bush, and other microfilm advocates played a major role in shaping the future of information technology.

---

[5] Yates, 1989.

As the cost of paper, printing, and transportation dropped in the late 19th and early 20th centuries, information dissemination and the profession of librarianship grew explosively. Library associations were formed. The Dewey Decimal and Library of Congress classification systems were developed. Thousands of relatively poorly funded public libraries sprang up to serve local demand in the United States. In Europe, government-funded libraries were established to serve scientists and other specialists in medicine and the humanities. This difference led to different approaches to technology development on either side of the Atlantic.

In the U.S., library management and the training of thousands of librarians took precedence over technology development and the needs of specialists. Public libraries adopted the simple but inflexible Dewey Decimal classification system. The pragmatic focus of libraries and emerging library schools meant that technology development was left to industry. Research into the indexing, cataloging, and retrieval of information and the physical objects containing it was variously referred to as bibliography, documentation, and documentalism.

In contrast, the well-funded European special libraries elicited sophisticated reader demands and pressure for libraries to share resources, which promoted interest in technology and information management. The Belgian Paul Otlet obtained Melvyn Dewey's permission to create an extended version of his classification system to support what we would today call hypertext links. Otlet agreed not to implement his Universal Decimal Classification (UDC) in English for a time, an early example of a legal constraint on technology development. Nevertheless, UDC is still in use in some places.

In 1926, the Carnegie Foundation dropped a bombshell by endowing the Graduate Library School (GLS) at the University of Chicago to focus solely on research. For two decades Chicago was the only university granting Ph.D.s in library studies. GLS positioned itself in the humanities and social sciences, with research into the history of publishing, typography, and other topics.[6] *An Introduction to Library Science* (Butler, 1933), the dominant library research textbook for forty years, was written at Chicago. *It did not mention information technology at all.* Library science was shaped by the prestigious GLS program until well into the computer era and human-tool interaction was not among its major concerns. Documentalists, researchers who did focus on technology, were concentrated in industry and government agencies.

Burke (2007) summarized the early history, with its emphasis on training librarians and other specialists:

> Most information professionals … were focusing on providing information to specialists as quickly as possible. The terms used by contemporary specialists appeared to be satisfactory for many indexing tasks and there seemed no need for systems based on comprehen-

---

[6]   Buckland, 1998.

sive and intellectually pleasing classification schemes. The goal of creating tools useful to non-specialists was, at best, of secondary importance.

My account emphasizes the points at which computer technologies came into what might be called "non-specialist use." This early history of information management is significant, however, because the web and declining digital storage costs have made it evident that everyone will soon become their own information manager, just as we are all now telephone operators. But I am getting ahead of our story. This chapter concludes with accounts of two individuals who, in different ways, shaped the history of information research and development.

2.2.1   PAUL OTLET AND THE MUNDANEUM

Like his contemporary H. G. Wells, Otlet envisioned a vast network of information. But unlike Wells, Otlet and his collaborators built one. Otlet established a commercial research service around facts that he had been cataloging on index cards since the late 19th century. In 1919 the Belgian government financed the effort, which moved to a record center called the Mundaneum. By 1934, 15 million index cards and millions of images were organized and linked or cross-referenced using UDC. Curtailed by the Depression and damaged during World War II, the work was largely forgotten. It was not cited by developers of the metaphorically identical Xerox Notecards, an influential hypertext system of the 1980s.

Technological innovation continued in Europe with the development of mechanical systems of remarkable ingenuity. Features included the use of photoreceptors to detect light passing through holes in index cards positioned to represent different terms, enabling rapid retrieval of items on specific topics.[7] These innovations inspired the work of Vannevar Bush, a well-known American scientist and research manager.

Vannevar Bush and Microfilm Machines

MIT professor Vannevar Bush was one of the most influential scientists in American history. He advised Presidents Franklin Roosevelt and Harry Truman, served as director of the Office of Scientific Research and Development, and was president of the Carnegie Institute.

Bush is remembered today for "As We May Think," his 1945 *Atlantic Monthly* essay. It described the memex, a hypothetical microfilm-based, electromechanical information-processing machine. The memex was to be a personal workstation that enabled a professional to quickly index and retrieve documents or pictures and create hypertext-like associations among them. The essay, excerpted below, inspired computer engineers and computer scientists who made major contributions to HCI in the 1960s and beyond.

---

[7]   Buckland, 2009.

What's not well known is that Bush wrote the core of his essay in the early 1930s, after which, shrouded in secrecy, he spent two decades and unprecedented resources on the design and construction of several machines that comprised a subset of memex features. None were successful. The details are recounted in Colin Burke's comprehensive *Information and Secrecy: Vannevar Bush, Ultra, and the Other Memex.*

Microfilm—photographic miniaturization—had qualities that attracted Bush, as they had Otlet. Microfilm was light, could be easily transported, and was as easy to duplicate as paper records (Xerox photocopiers did not appear until 1959). The cost of handling film was brought down by technology created for the moving picture industry. Barcode-like patterns of small holes could be punched on a film and read very quickly by passing the film between light beams and photoreceptors. Microfilm was tremendously efficient as a storage medium. Memory based on relays or vacuum tubes would never be competitive, and magnetic memory, when it eventually arrived, was less versatile and far more expensive. It is easy today to overlook the compelling case that existed for basing information systems on microfilm.

Bush's machines failed because of overly ambitious compression and speed goals and patent disputes, but ultimately most critical was that Bush was unaware of decades of research on classification systems. American documentalists had been active, albeit not well funded. In 1937, the American Documentation Institute (ADI) was formed, a predecessor of the Association for Information Science and Technology (ASIS&T). Had he worked with them, Bush, an electrical engineer by training, could have avoided the fatal assumption that small sets of useful indexing terms would easily be defined and agreed upon. Metadata design was a research challenge then, and still is.

Bush described libraries and the public as potential users, but his machines cost far too much for that. He focused on the FBI and CIA as customers, as well as military uses of cryptography and information retrieval. Despite the classified nature of this work, through his academic and government positions, his writings, the vast resources he commandeered, and the scores of brilliant engineers he enlisted to work on microfilm projects, Bush exerted influence for two decades, well into the computer era.

Bush's vision emphasized both associative linking of information sources and discretionary use:

Associative indexing, the basic idea of which is a provision whereby any item may be caused at will to select immediately and automatically another. This is the essential feature of the memex... Any item can be joined into numerous trails... New forms of encyclopedias will appear, ready made with a mesh of associative trails [which a user could extend]
...

The lawyer has at his touch the associated opinions and decisions of his whole experience and of the experience of friends and authorities. The patent attorney has on call the millions of issued patents, with familiar trails to every point of his client's interest. The physician, puzzled by a patient's reactions, strikes the trail established in studying an earlier similar case and runs rapidly through analogous case histories, with side references to the classics for the pertinent anatomy and histology. The chemist, struggling with the synthesis of an organic compound, has all the chemical literature before him in his laboratory, with trails following the analogies of compounds and side trails to their physical and chemical behavior.

The historian, with a vast chronological account of a people, parallels it with a skip trail, which stops only on the salient items, and can follow at any time contemporary trails which lead him all over civilization at a particular epoch. There is a new profession of trail blazers, those who find delight in the task of establishing useful trails through the enormous mass of the common record.

Bush knew that the memex was unrealistic. None of his many projects included the "essential" associative linking. Nevertheless, his descriptions of discretionary hands-on use of powerful machines by professionals was inspirational. His vision was realized 50 years later, built on technologies undreamt of in the 1930s and 1940s. Bush did not initially support computer development—their slow, bulky, and expensive information storage was clearly inferior to microfilm.

Bearers of information. Wells, Otlet, and Bush focused on index cards and microfilm, but the input and output technologies that would dominate the first decades of computing were already present: paper tape and especially punch cards. Early computers held one program at a time, so each job required loading a program and any required data. The program/data distinction was present centuries earlier. Paper tape with punched holes was used to program mechanical looms in the 18th century and punch cards to store information in the mid-19th century. Charles Babbage proposed punched "number cards" for the control of a "difference engine," or calculating engine. Herman Hollerith, inspired by cards on which train conductors punched a set of features describing each passenger, devised punch cards that were used for the 1890 U.S. Census. By the 1930s, millions were being produced daily. Their most successful descendant was the 80-column "IBM card," used to input programs and data and onto which output could be punched.

A typical computer center from the 1950s through the mid-1980s had keypunches, into which blank cards were loaded for entering program instructions and data; a card reader, into which cards were loaded; the computer itself, with switches and buttons for inputting commands; and a teletype that produced continuous paper output with perforations to separate, or "burst," the pages. It could also have a computer-controlled card punch for outputting data and a drive for magnetic tape, a storage medium that like a hard drive only the computer read. There might be a card sorter, a visually arresting machine that sorted cards by shooting them along a track and dropping them into different bins based on the holes in a specified column. Paper tape with punched holes was used into the 1970s. By the late 1980s, interactive terminals and disk drives had displaced much of this, with magnetic tape mainly used for backup.

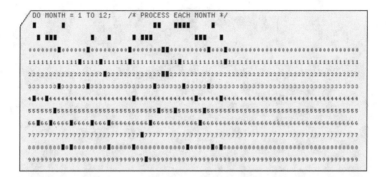

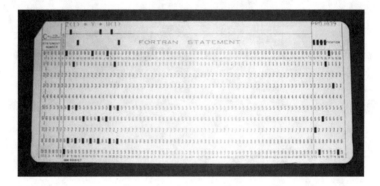

Figure 2.1: Two punch cards, long the most common medium for interacting with computers. Photo credits: Arnold Reinhold and Jeff Barr CC BY-SA 2.5.

CHAPTER 3

1945–1955: Managing Vacuum Tubes

World War II changed everything. Until then, most government research funding was managed by the U.S. Department of Agriculture. The war brought unprecedented investment in science and technology, culminating in the atomic bomb. This development showed that huge sums could be found for *academic or industrial research that addressed national goals*. Research expectations and strategies would never be the same.

Sophisticated electronic computation machines built before and during World War II were designed for specific purposes, such as solving equations or breaking codes. Each of the expensive cryptographic machines that helped win the war was designed to attack a specific encryption device. Whenever the enemy changed devices, a new machine was needed. This spurred interest in developing general-purpose computational devices. Wartime improvements in vacuum tubes and other technologies made this more feasible, and their deployment brought HCI into the foreground.

When engineers and mathematicians emerged from military and government laboratories and secret project rooms on university campuses, the public became aware of some breakthroughs. Development of ENIAC, arguably the first general-purpose computer, had begun in secret during the war; the "giant brain" was revealed publicly only when it was completed in 1946. Its first use was not publicized: calculations to support hydrogen bomb development. ENIAC stood 8–10 feet high, occupied about 1,800 square feet, and consumed as much energy as a small town. It provided far less computation and memory than you can slip into your pocket and run on a small battery today.

Memory was inordinately expensive. Even the largest computers of the time had little memory, so they were used for computation and not for symbolic representation or information processing. The HCI focus was to reduce operator burden, enabling a person to replace or reset vacuum tubes more quickly and load stored-program computers from tape rather than by manually attaching cables and setting switches. Such "knobs and dials" human factors improvements enabled one computer operator to accomplish work that had previously required a team.

Libraries installed simple microfilm readers to assist with information retrieval as publication of scholarly and popular material soared, but interest in technology was otherwise limited. The GLS orientation still dominated, focused on librarianship, social science, and historical research. Independently, the foundation of information science was coming into place, built on alliances that had been forged during the war among documentalists, electrical engineers, and mathematicians interested in communication and information management. These included Vannevar Bush and his collaborators Claude Shannon and Warren Weaver, co-authors in 1949 of a seminal work on information theory called communication theory at the time, and Ralph Shaw, a prominent American documentalist. The division between the two camps widened. Prior to the war, the technology-oriented ADI included librarians and support for systems that spanned humanities and sciences; during the war and thereafter, ADI focused on government and "big science."

3.1   THREE ROLES IN EARLY COMPUTING

Early computer projects employed people in three roles: management, programming, and operation. Managers, who often had backgrounds in science or engineering, specified the programs to be written; oversaw the design, development, and operation; and handled the output. Mathematically adept programmers decomposed tasks into components that a computer could manage. The first professional programming team comprised six women working on ENIAC. At the time, they were called "computers." They also debugged programs, sometimes by crawling into the giant machine to locate problems. In addition, a small army of operators was needed. Once written, a program could take days to load by setting switches, positioning dials, and connecting cables. Despite innovations that boosted reliability, such as operating vacuum tubes at lower power to increase life span and designing visible indicators of tube failure, ENIAC was often stopped so that failed vacuum tubes could be located and replaced. Vacuum tubes were reportedly wheeled around in shopping carts.

Eventually, each occupation—management and systems analysis, programming, and operation—became a major focus of HCI research, centered respectively in (management) information systems, computer science, and human factors. Significant HCI contributions from information science and other disciplines awaited greater processing power and sharply reduced prices for digital memory.

3.1.1   GRACE HOPPER: LIBERATING COMPUTER USERS

As computers became more reliable and capable, programming became a central activity. Computer languages, compilers, and constructs such as subroutines facilitated "programmer-computer interaction." Grace Hopper was a pioneer in all of these areas. She described her goal as freeing mathematicians to do mathematics.[8] It was echoed years later in the HCI goal of freeing users to do their work. In the early 1950s, mathematicians *were* the users! Hopper's work led to the development of COBOL, an English-like programming language for business systems that became and may still be the programming language in most widespread use in the world. At IBM, John Backus led the development of FORTRAN, also a highly successful English-like programming language, designed for scientific computing and released in 1957.

Just as HCI professionals often feel marginalized by software developers, Hopper's pioneering accomplishments in HCI were arguably undervalued by other computer scientists, although

---

[8]   Hopper, 1952; Sammet, 1992.

she received recognition through the annual Grace Hopper Celebration of Women in Computing, initiated in 1994.

CHAPTER 4

1955–1965: Transistors, New Vistas

Early forecasts that the world would need only a few computers reflected the limitations of vacuum tubes. This changed when solid-state computers became available commercially in 1958. Computers were still used primarily for scientific and engineering tasks, but they were now reliable enough not to require a large staff to maintain one computer. As loading and running programs became more routine, operators with less engineering knowledge would be sufficient once more intuitive interfaces were developed. Although transistor-based computers were still very expensive and had limited capabilities, researchers could envision the previously unimaginable possibility of computers operated by people with no technical training.

The Soviet Union's launch of the Sputnik satellite in 1957 challenged the West to invest in science and technology. The development of lighter and more capable computers was an integral part of the well-funded program that put men on the moon 12 years later.

4.1   SUPPORTING OPERATORS: THE FIRST FORMAL HCI STUDIES

"In the beginning, the computer was so costly that it had to be kept gainfully occupied for every second; people were almost slaves to feed it."

—Brian Shackel (1997)

Almost all computer use in the late 1950s and early 1960s involved programs and data that were read in from cards, paper tape, or magnetic tape. A program then ran without interruption until it terminated, producing printed, punched, or tape output. This "batch processing" restricted human interaction to operating the hardware, programming, and using the output. Of these, the only job involving hands-on computer use was the least challenging and lowest paying, the computer operator. Programs were typically written on paper and keypunched onto cards or tape.

Computer operators loaded and unloaded cards and tapes, set switches, pushed buttons, read lights, loaded and burst printer paper, and put printouts into distribution bins. Operators interacted directly with the system via a teletype: typed commands interleaved with computer responses and status messages were printed on paper that scrolled up one line at a time. Eventually, printers yielded to "glass tty's" (glass teletypes), also called cathode ray tubes (CRTs), and visual display units or terminals (VDUs/VDTs). These displays also scrolled commands and computer responses one line at a time. The price of a monochrome terminal that could only display alphanumeric characters was $50,000 in today's dollars, a small fraction of the cost of the computer. A large computer might have one or more consoles. Programmers did not use interactive consoles until later. Although the capabilities were far less than that of a tablet today, the charge to use an IBM 650 was $7500 an hour.[9]

Improving the design of buttons, switches, and displays was a natural extension of human factors or ergonomics. In 1959, British ergonomist Brian Shackel published the first HCI paper, "Ergonomics for a Computer," followed in 1962 by his "Ergonomics in the Design of a Large Digital Computer Console." These described console redesign for analog and digital computers called the EMIac and EMIdec 2400, the latter being the largest computer at the time of the study (Shackel, 1997).

In the U.S., in 1956, aviation psychologists created the Human Engineering Society, focused on improving skilled performance through greater efficiency, fewer errors, and better training. The next year, it adopted the more elegant title Human Factors Society and, in 1958, it initiated the journal *Human Factors*. Sid Smith's (1963) "Man–Computer Information Transfer" marked the start of his long career in the human factors of computing.

4.2   VISIONS AND DEMONSTRATIONS

As transistors replaced vacuum tubes, a wave of imaginative writing, conceptual innovation, and prototype building swept through the research community. Some of the terms and language quoted below are dated, notably in the use of male generics, but many of the key concepts still resonate.

4.2.1   J.C.R. LICKLIDER AT MIT, BBN, AND ARPA

A psychologist and engineer, Licklider played a dual role in advancing computer science and HCI.[10] He wrote the influential essays described in the preamble and supported important research projects as a vice president at Bolt, Beranek and Newman (BBN) from 1957 to 1962, as head of the Information Processing Techniques Office (IPTO) and director of Behavioral Sciences Command and Control Research at ARPA (sometimes called DARPA) from 1962 to 1964, and later as director of MIT's Project MAC.

Among the influential researchers employed by BBN to work on government-funded computer-related projects were John Seely Brown, Richard Pew, and MIT faculty John McCarthy and Marvin Minsky. IPTO funding created the first computer science departments and established artificial intelligence (AI) as a discipline in the 1960s. It is best known for the internet predecessor ARPANET, which operated from 1969 to 1985.

As noted in the preamble, Licklider (1960) outlined a vision of human-machine symbiosis: "There are many human-machine systems. At present, however, there are no human-computer symbioses." The computer, he wrote, was "a fast information-retrieval and data-processing machine"

---

[9]   Markoff, 2015.
[10]   Waldrop, 2001.

destined for a larger role: "One of the main aims of human–computer symbiosis is to bring the computing machine effectively into the formulative parts of technical problems."

This would require rapid, real-time interaction, which was not supported by the prevailing batch systems that ran one program at a time from start to finish without intervention. In 1962, Licklider and Wesley Clark identified requirements of a system for "on-line human–computer communication" that they felt were ripe for development: time-sharing of a computer among users; electronic input–output surfaces to display and communicate symbolic and pictorial information; interactive, real-time support for programming and information processing; large-scale information storage and retrieval systems; and facilitation of human cooperation. They were right: The ensuing decades of HCI work filled in this outline. They also foresaw that speech recognition and natural language understanding would be more difficult to achieve.

4.2.2   JOHN MCCARTHY, CHRISTOPHER STRACHEY, WESLEY CLARK

McCarthy and Strachey worked on time-sharing, which made interactive computing possible.[11] Apart from a few researchers who had access to computers built with spare-no-expense military funding, computer use was too expensive to support exclusive individual access. Time-sharing allowed several simultaneous users (and later dozens) to work at terminals cabled to a single computer. Languages were developed to facilitate control and programming of time-sharing systems (e.g., JOSS in 1964).

Wesley Clark was instrumental in building the TX-0 and TX-2 at MIT's Lincoln Laboratory. These machines, which cost on the order of US$10 million apiece, demonstrated time-sharing and other innovative concepts, and helped establish the Boston area as a center for computer research. The TX-2 was the most powerful and capable computer in the world at the time. It was much less powerful than a smartphone is today. Buxton (2006) includes a recording of Clark and Ivan Sutherland discussing this era in 2005.

4.2.3   IVAN SUTHERLAND AND COMPUTER GRAPHICS

Sutherland's 1963 Ph.D. thesis may be the most influential document in the history of HCI. His Sketchpad system, built on the TX-2 to make computers "more approachable," launched computer graphics, which had a decisive impact on HCI 20 years later. Sutherland's thesis and videos of him demoing Sketchpad are available online.[12]

Sutherland demonstrated iconic representations of software constraints, object-oriented programming concepts, and the copying, moving, and deleting of hierarchically organized objects. He explored novel interaction techniques, such as picture construction using a light pen. He facilitated

[11]   Fano and Corbato, 1966.
[12]   Thesis: http://www.cl.cam.ac.uk/TechReports/UCAM-CL-TR-574.pdf. Videos: search YouTube for "Ivan Sutherland Sketchpad."

visualization by separating the coordinate system used to define a picture from the one used to display it and demonstrated animated graphics, noting the potential for digitally rendered cartoons 20 years before *Toy Story*. His frank account enabled others to make rapid progress—when engineers found Sketchpad too limited for computer-assisted design (CAD), he called the trial a "big flop." He identified the obstacles and soon CAD was thriving.

In 1964, Sutherland succeeded J.C.R. Licklider as the director of IPTO. Among those he funded was Douglas Engelbart at the Stanford Research Institute (SRI).

4.2.4   DOUGLAS ENGELBART: AUGMENTING HUMAN INTELLECT

In 1962, Engelbart published "Augmenting Human Intellect: A Conceptual Framework." In the years that followed, he built systems that made astonishing strides toward realizing his vision. He also supported and inspired engineers and programmers who made major contributions.

Echoing Bush and Licklider, Engelbart saw the potential for computers to become congenial tools that people would choose to use interactively:

> By "augmenting human intellect" we mean increasing the capability of a man to approach a complex problem situation, to gain comprehension to suit his particular needs, and to derive solutions to problems... By "complex situations" we include the professional problems of diplomats, executives, social scientists, life scientists, physical scientists, attorneys, designers... We refer to a way of life in an integrated domain where hunches, cut-and-try, intangibles, and the human "feel for a situation" usefully co-exist with powerful concepts, streamlined terminology and notation, sophisticated methods, and high-powered electronic aids.

Engelbart used ARPA funding to develop and integrate an extraordinary set of prototype applications into his NLS system. He conceptualized and implemented the foundations of word processing, invented and refined input devices including the mouse and a multi-key control box, and built multi-display environments that integrated text, graphics, hypertext, and video. These unparalleled advances were demonstrated in a sensational 90-minute live event at the 1968 Fall Joint Computer Conference in San Francisco.[13] This event, later called "The Mother of all Demos," was a factor in drawing interactive systems researchers to the U.S. West Coast. It is available online.

Engelbart, an engineer, supported human factors testing to improve efficiency and reduce errors in skilled use arising from fatigue and stress, but he was less concerned with the initial experience. Engelbart felt that people should be willing to undergo training to tackle a difficult interface if it would deliver great power once mastered. Unfortunately, difficulty with initial use was a factor in Engelbart's subsequent loss of funding. The 1968 demo was in one respect a success disaster:

---

[13]   http://sloan.stanford.edu/MouseSite/1968Demo.html.

DARPA installed NLS in its offices and found it too difficult.[14] Years later, the question "Is it more important to optimize for skilled use or initial use?" was actively debated by CHI researchers, and it still resurfaces.

4.2.5   TED NELSON'S VISION OF INTERCONNECTEDNESS

In 1960, Ted Nelson, then a graduate student in sociology, founded Project Xanadu and three years later coined the term "hypertext" in describing the goal, an easily used computer network. In 1965, he published an influential paper titled "A File Structure for the Complex, the Changing and the Indeterminate." Nelson wrote stirring calls for systems that would democratize computing through a highly interconnected, extensible network of digital objects (e.g., Nelson, 1973). Xanadu was never fully realized, and Nelson did not consider the early world wide web to be a realization of his vision, but subsequent, lightweight technologies such as weblogs, wikis, collaborative tagging, and search enable many of the activities he envisioned.

Later, Nelson (1996) anticipated intellectual property issues that would arise in digital domains. He coined the term "micropayment," again drawing attention to possibilities that were subsequently realized in different ways.

4.3   FROM DOCUMENTALISM TO INFORMATION SCIENCE

The late 1950s saw the final major investments in microfilm and other pre-digital systems. The most ambitious were Vannevar Bush's final military and intelligence projects.[15] Some documentalists saw that the declining cost of digital memory would enable computation engines to become information-processing machines. As mathematicians and engineers engaged in technology development, their initiatives had few ties to contemporary librarianship or to the humanities orientation of library schools. The conceptual evolution was incremental, but institutional change came swiftly in some cases, organized around a new banner: information science.

Merriam Webster dates the term "information science" to 1960. Conferences on "Training Science Information Specialists" organized by Dorothy Crosland at the Georgia Institute of Technology in 1961 and 1962 are credited with shifting the focus from information as a technology to information as an incipient science. In 1963, chemist-turned-documentalist Jason Farradane taught the first information science courses at City University, London. Much earlier, the profession of chemistry had taken the lead in organizing its literature systematically. Another chemist-turned-documentalist, Allen Kent, was central to an information science initiative at the University of Pittsburgh.[16] In the early 1960s, Anthony Debons, a psychologist and friend of J. C. R. Licklider, organized a series of NATO-sponsored congresses at the University of Pittsburgh.

---

[14]   Bardini, 2000.
[15]   Burke, 1994.
[16]   Aspray, 1999.

Guided by Douglas Engelbart, these meetings centered on people and on how technology could augment their activities. In 1964, the Graduate Library School at the University of Pittsburgh became the Graduate School of Library and Information Sciences, and the "School of Information Science" at Georgia Tech opened, developed by Crosland with National Science Foundation (NSF) backing and initially employing one full-time faculty member.[17]

4.4   CONCLUSION: VISIONS, DEMOS, AND WIDESPREAD USE

Progress in HCI can be reflected in inspiring visions, conceptual advances that enable aspects of those visions to be demonstrated in working prototypes, and the evolution of design and application. The engine that drove the visions and enabled many to be realized and eventually widely deployed was the relentless advance of hardware that produced devices millions of times more powerful than the far more expensive systems of the pioneers.

At the conceptual level, the basic foundation for today's graphical user interfaces was in place by 1965. As Sutherland worked on the custom-built US$10 million TX-2 at MIT, a breakthrough occurred nearby: the Digital Equipment Corporation PDP-1 minicomputer—"truly a computer with which an individual could interact."[18] First appearing in 1960, the PDP-1 cost $1 million and came with a CRT display, keyboard, light pen, and paper tape reader. Although it was more powerful than the TX-2, today it is difficult to appreciate how little a million dollars bought. It had about the capability of a $2,000 Radio Shack TRS 80 in 1977—and it required full-time technical and programming support![19] Today you can buy computers for $50 that far outperform these predecessors. Nevertheless, the PDP-1 and its descendants became the machines of choice for computer-savvy researchers.

Licklider's "man-computer symbiosis," Engelbart's "augmented human intellect," and Nelson's "conceptual framework for man-machine everything" described a world that did not exist: a world in which attorneys, doctors, chemists, and designers chose to be hands-on users of computers. For some time to come, the reality would be that most hands-on use was by computer operators engaged in routine, nondiscretionary tasks. As for the visions, 50 years later, many of the capabilities are taken for granted, some are just being realized, and a few remain elusive.

---

[17]   Smith, 2004.
[18]   Pew, 2003.
[19]   Pew, 2003.

CHAPTER 5

1965–1980: HCI Prior to Personal Computing

Business computing took off with the arrival and corporate embrace of mainframe computers in the late 1960s.[20] Control Data Corporation launched the transistor-based 6000 series computers in 1964. In 1965, commercial computers based on integrated circuits arrived with the IBM System/360. These systems, later called mainframes to distinguish them from minicomputers, firmly established computing in the business realm. Each of the three computing roles—operation, management, and programming—became a significant profession in this period.

Operators still interacted directly with computers for routine maintenance and operation. As time-sharing spread, hands-on use expanded to include data entry and other repetitive tasks. Managers and systems analysts oversaw hardware acquisition, software development, operation, and the use of output. Managers who relied on printed output and reports were called "computer users," although they did not interact directly with the computers.

Few programmers were direct users until late in this period. Most prepared flowcharts and wrote programs on paper forms. Keypunch operators then punched the program instructions onto cards. These were sent to computer centers for computer operators to load and run. Printouts and other output were picked up later. Some programmers used computers directly when they could, but unless they worked in research centers, the cost of computer time generally dictated a more efficient division of labor.

5.1 HUMAN FACTORS AND ERGONOMICS EMBRACE COMPUTER OPERATION

As business turned to computing, the high cost of hardware generated intense interest in efficiency, a hallmark of human factors since the era of Taylor and the Gilbreths. In 1970, Brian Shackel founded the Human Sciences and Advanced Technology (HUSAT) ergonomics research center at Loughborough University, emphasizing HCI. In the U.S., Sid Smith and colleagues worked on input and output issues, such as information representation on displays and computer-generated speech (Smith et al., 1965; Smith and Goodwin, 1970). The Computer Systems Technical Group (CSTG) formed in 1972.

---

[20] The first business computer was the Lyons Electronic Office, or LEO, in the 1950s and early 1960s. Although perhaps not profitable, it is a remarkable instance of farsighted business execution. See the Wikipedia entry for Leo (computer).

HCI articles appeared in the *Human Factors* journal and, in 1969, the *International Journal of Man-Machine Studies* (*IJMMS*) began publication. The first widely read HCI book was James Martin's (1973) *Design of Man–Computer Dialogues*. His comprehensive survey of interfaces for computer operators and data entry personnel began with an arresting opening chapter that described a world in transition. Extrapolating from declining hardware prices, he wrote:

> The terminal or console operator, instead of being a peripheral consideration, will become the tail that wags the whole dog… The computer industry will be forced to become increasingly concerned with the usage of people, rather than with the computer's intestines.

In the mid-1970s, the U.S. government agencies responsible for agriculture and social security initiated large-scale data processing projects. Although unsuccessful, these efforts led to methodological innovations in the use of style guides, usability labs, prototyping, and task analysis. [21]

In 1980, three significant human factors and ergonomics (HF&E) books were published: two on VDT design and one on general guidelines. [22] Drafts of German work on VDT standards, made public in 1981, provided an economic incentive to design for human capabilities by threatening to ban noncompliant products. Later that year, an American National Standards Institute (ANSI} "office and text systems" working group formed.

Human factors had other concerns in this period. The Three Mile Island nuclear power disaster of 1979 was attributed to the design of displays monitoring non-digital electro-mechanical systems and the training and reactions of their operators. Months later, the National Academy of Sciences (NAS) created a Committee on Human Factors at the urging of the armed forces. Nor were the human factors issues related to the Space Shuttle Challenger launch disaster six years later attributed to digital technology. Computation was still a niche activity, but activity increased: CSTG became the largest technical group of the HFS.

5.2   INFORMATION SYSTEMS ADDRESSES THE MANAGEMENT OF COMPUTING

Companies acquired expensive business computers to address major organizational concerns. Even when the principal concern was simply to appear modern (Greenbaum, 1979), the desire to show benefit from a multi-million dollar investment chained some managers to a computer almost as firmly as it did the operator and data entry "slaves." They were expected to manage any employee resistance to using a system and to make use of the output.

In 1967, the journal *Management Science* initiated a column titled "Information Systems in Management Science." Early definitions of "information systems" included "an integrated man/machine system for providing information to support the operation, management, and decision-mak-

[21]   Pew, 2003.
[22]   **VDT standards:** Cakir, et al., 1980; Grandjean and Vigliani, 1980; **general guidelines:** Damodaran et al., 1980.

ing functions in an organization" and "the effective design, delivery and use of information systems in organizations."[23] In 1968, a management information systems center and degree program was established at the University of Minnesota. It initiated influential research streams and in 1977 launched *MIS Quarterly*, which has been the field's leading journal. MIS focused on specific tasks in organizational settings, emphasizing both general theory and precise measurement, a challenging combination.

A historical survey by Banker and Kaufmann (2004) identifies HCI as one of five major IS research streams; the others focus on the economics of information, the economics of information technology, strategy, and decision making. They date the HCI stream back to a paper by Ackoff (1967) on challenges in handling computer-generated information. Some MIS research overlapped HF&E concerns with hands-on operator issues in data entry and error message design, but most early HCI work in information systems dealt with the users of information, typically managers. Research included the design of printed reports. The drive for theory led to a strong focus on cognitive styles: individual differences in how people (especially managers) perceive and process information. MIS articles on HCI appeared in the human factors–oriented *IJMMS* as well as in management journals.

Enid Mumford and others developed sociotechnical approaches to system design in this period in response to user difficulties and resistance.[24] The methods included educating representative workers about technological possibilities and involving them in design, in part to increase employee acceptance of the system being developed. A cooperative design approach led by Kristin Nygaard emerged from the Scandinavian trade union movement. It focused on empowering future hands-on users of a system being developed.[25] Influential, sophisticated views of the complex social and organizational dynamics around system adoption and use included Rob Kling's "Social Analyses of Computing: Theoretical Perspectives in Recent Empirical Research" (1980) and Lynne Markus's "Power, Politics, and MIS Implementation" (1983).

5.3 PROGRAMMING: SUBJECT OF STUDY, SOURCE OF CHANGE

Programmers who were not hands-on users nevertheless interacted with computers. More than 1,000 research papers on variables that affected programming performance were published in the 1960s and 1970s.[26] Most examined programmer behavior in isolation, independent of organizational context. Influential reviews included Gerald Weinberg's landmark *The Psychology of Computer Programming* in 1971, Ben Shneiderman's *Software Psychology: Human Factors in Computer and*

---

[23] Davis (1974) and Keen (1980), both quoted in Zhang et al. (2004).
[24] Mumford, 1971, 1976; Bjørn-Andersen and Hedberg, 1977.
[25] Nygaard, 1977.
[26] Baecker and Buxton, 1987.

Information Systems in 1980, and Beau Sheil's 1981 review of studies of programming notation (conditionals, control flow, data types), practices (flowcharting, indenting, variable naming, commenting), and tasks (learning, coding, debugging).

Software developers changed the field through invention. In 1970, Xerox Palo Alto Research Center (PARC) was created to develop new computer hardware, programming languages, and programming environments. It attracted researchers and system builders from the laboratories of Engelbart and Sutherland. In 1971, Allen Newell of Carnegie Mellon University proposed a project to PARC on the psychology of cognitive behavior, writing that "central to the activities of computing—programming, debugging, etc.—are tasks that appear to be within the scope of this emerging theory." It was launched in 1974.[27]

HUSAT and PARC were founded in 1970 with broad charters. HUSAT focused on ergonomics, anchored in the tradition of nondiscretionary use, one component of which was the human factors of computing. PARC focused on computing, anchored in visions of discretionary use, one component of which was also the human factors of computing. Researchers at PARC, influenced by cognitive psychology, extended the primarily perceptual-motor focus of human factors to higher-level mental processes, whereas HUSAT, influenced by sociotechnical design, extended human factors by considering organizational factors.

5.4   COMPUTER SCIENCE: A NEW DISCIPLINE

The first university computer science departments formed in the mid-1960s. Their orientation depended on their origin: some branched from engineering, others from mathematics. Computer graphics was an engineering specialization of particular relevance to HCI. Applied mathematics provided many of the early researchers in AI, which has interacted with HCI in complex ways. The next two sections cover these influences.

Almost all early machines were funded by branches of the military, with no concern for their cost. Technical success was the sole evaluation criterion.[28] Directed by Licklider, Sutherland, and their successors, ARPA played a major role. The high cost concentrated researchers in a few centers that bore little resemblance to the business computing environments of that era. Users and their needs differed: technically savvy hands-on users in research settings did not press for the low-level interface efficiency enhancements important to business.

The computer graphics and AI perspectives in these research centers differed from those of the HCI researchers who focused on less expensive, more widely deployed systems. Computer graphics and AI required processing power—hardware advances led to *declining costs for the same high level of computation.* For HCI research, hardware advances led to *greater computing capability at*

---

[27] Card and Moran, 1986.

[28] Norberg and O'Neill, 1996. This was also stressed by Wesley Clark when I interviewed him in 2010.

the same low price. This difference would diminish in years to come, when widely available machines could support graphical interfaces and some AI programs. Nevertheless, some computer scientists focused on interaction between 1965 and 1980, often influenced by the central role of discretionary use in the writings of Bush, Licklider, and Engelbart.

The University of Illinois at Urbana-Champaign was the first university to build and own a computer, on which PLATO, the first computer-assisted instruction program, was developed in the 1960s. In the 1970s, PLATO introduced novel email and group communication features and a terminal network that by the mid-1980s provided instruction in about 100 subjects to students at dial-up terminals around the world. Innovation in communication applications continued with the popular Eudora email system (1988) and Mosaic browser (1993).

5.4.1   COMPUTER GRAPHICS: REALISM AND INTERACTION

In 1968, Sutherland joined the computer graphics laboratory at the University of Utah, where David Evans had established a computer science department in 1965. Most graphics work at the time was on PDP-1 and PDP-7 minicomputers. The list price of a high-resolution display was more than US$100,000 in today's dollars. The machines were in principle capable of multi-tasking, but in practice most graphics programs required all of a processor's cycles.

In 1969, J. C. R. Licklider keynoted at the first Canadian Man-Computer Communications conference. Most papers focused on computer graphics, and the conference series was rechristened Graphics Interface in 1982, but some had no graphics and many were unequivocally HCI. For example, a 1975 paper was titled "Human Factors in Interactive Computer Graphics." That same year the panel "Human Factors Considerations in Computer Graphics" convened at the second SIGGRAPH conference.

The Xerox Alto was a step toward realizing Alan Kay's vision of computation as a medium for personal computing when it arrived in 1973.[29] It was not powerful enough to support high-end graphics research, but it included polished versions of graphical interface features that Engelbart prototyped five years earlier. Less expensive than the PDP-1 but too costly for people outside Xerox PARC, the Alto was not widely marketed. It signaled the approach of affordable, interactive, personal machines with a graphical interface, but for a decade, few people outside PARC could access one. Once in Palo Alto, few researchers left.

Computer graphics researchers had to choose: high-end graphics, or more primitive features that could run on widely affordable machines. William Newman (1973) co-author of the influential *Principles of Interactive Computer Graphics*, described the shift: "Everything changed—the Computer Graphics community got interested in realism, I remained interested in interaction, and I eventually found myself doing HCI" (personal communication). Other graphics researchers

---

[29] Kay, 1969; Kay and Goldberg, 1977.

whose focus shifted to broader interaction issues included Ron Baecker and Jim Foley. Foley and Wallace (1974) identified requirements for designing "interactive graphics systems whose aim is good symbiosis between man and machine." Eighteen papers in the first SIGGRAPH conference in 1974 had the words "interactive" or "interaction" in their titles. A decade later, none did. SIGGRAPH focused on photorealism; researchers and developers whose primary interest was interaction shifted to SIGCHI.

Prior to the Alto, HCI research had focused on increasing the efficiency of people who were trained to get past initial difficulties. At Xerox, Larry Tesler and Tim Mott took the step of considering how a graphical interface could best serve users who had no training or technical background. By early 1974, they had developed the Gypsy text editor. Gypsy and the concurrently developed Bravo editor preceded Microsoft Word; Charles Simonyi led the development of Bravo for Xerox and Word for Microsoft.[30]

In 1976, SIGGRAPH sponsored a two-day workshop titled "User Oriented Design of Interactive Graphics Systems." It was chaired by Sigfried Treu, the University of Pittsburgh's computer science department chair, and included Anthony Debons of Pittsburgh's influential information science program. Workshop participants who were later active in CHI included Jim Foley, William Newman, Ron Baecker, John Bennett, Phyllis Reisner, and Tom Moran. Licklider and Nicholas Negroponte presented vision papers. UODIGS'76 marked the end of the visionary period. Licklider (1976) saw it clearly:

> Interactive computer graphics appears likely to be one of the main forces that will bring computers directly into the lives of very large numbers of people during the next two or three decades. Truly user-oriented graphics of sufficient power to be useful to large numbers of people has not been widely affordable, but it will soon become so, and, when it does, the appropriateness and quality of the products offered will to a large extent determine the future of computers as intellectual aids and partners of people.

Despite the stature of the participants, the 150-page proceedings were unavailable to non-attendees and virtually never cited. The next ACM "user-oriented design" conference was held five years later, after which they became annual events. Application of graphics was not yet at hand; in 1980, HCI research remained focused on interaction that was driven by commands, forms, and full-page menus.

5.4.2   ARTIFICIAL INTELLIGENCE: WINTER FOLLOWS SUMMER

AI burst onto the scene in the late 1960s and early 1970s. Logically, AI and HCI are closely related. What are intelligent machines for, if not to interact with people? AI research influenced HCI: speech recognition and natural language are perennial HCI topics; expert, knowledge-based, adap-

---

[30]   Hiltzik, 1999.

tive, and mixed-initiative systems engaged some human factors and CHI researchers, as did HCI applications of production systems, neural networks, and fuzzy logic.

However, AI did not transform HCI. Some AI features made it into systems and applications, but predictions that powerful AI technologies would come into wide use were not borne out. AI did not become a major facet of the HCI research literature, and few AI researchers showed interest in HCI.

To understand how this transpired requires a brief review of AI history. The term "artificial intelligence" first appeared in a 1955 call by John McCarthy for a meeting on machine intelligence held at Dartmouth. The British logician and code breaker Alan Turing's (1950) prescient essay "Computing Machinery and Intelligence" attracted attention when it was reprinted in *The World of Mathematics* in 1956. (Also published in 1950 were Claude Shannon's "Programming a Computer for Playing Chess"and Isaac Asimov's thoughtful exploration of ethical issues in the context of science fiction stories *I, Robot*). Also in 1956, Newell and Simon outlined a logic theory machine, after which they focused on developing a general problem solver. The LISP programming language, devised to support AI, was formulated in 1958.[31]

Many AI pioneers were trained in mathematics and logic, fields that can be largely derived from a few axioms and a small set of rules. Mathematical ability is considered a high form of intelligence, even by non-mathematicians. AI researchers anticipated that machines that operate logically and tirelessly would make profound advances by applying a small set of rules to a limited number of objects. Early AI focuses included theorem-proving and games such as chess and go, which like math start with a small number of rules and a fixed set of objects. McCarthy (1988), who espoused predicate calculus as a foundation for AI, summed it up:

> As suggested by the term "artificial intelligence," we weren't considering human behavior except as a clue to possible effective ways of doing tasks. The only participants who studied human behavior were Newell and Simon. (The goal) was to get away from studying human behavior and consider the computer as a tool for solving certain classes of problems. Thus, AI was created as a branch of computer science and not as a branch of psychology.

Unfortunately, by ignoring psychology, mathematicians overlooked the complexity and inconsistency that mark human thought and social constructs. Underestimating the complexity of intelligence, they overestimated the prospects for creating it artificially. Hyperbolic predictions and AI were close companions from the start. In the summer of 1949, Alan Turing wrote in the *London Times*:

> I do not see why [the computer] should not enter any one of the fields normally covered by the human intellect, and eventually compete on equal terms. I do not think you can even draw the line about sonnets, though the comparison is perhaps a little bit unfair

---

[31]   McCarthy, 1960.

because a sonnet written by a machine will be better appreciated by another machine (Turing, 1949).

Optimistic forecasts by the 1956 Dartmouth workshop participants attracted considerable attention. When they collided with reality, a pattern was established that was to play out repeatedly. Hans Moravec (1998) wrote:

> In the 1950s, the pioneers of AI viewed computers as locomotives of thought, which might outperform humans in higher mental work as prodigiously as they outperformed them in arithmetic, if they were harnessed to the right programs… By 1960 the unspectacular performance of the first reasoning and translation programs had taken the bloom off the rose.

In 1960, the managers of MIT's Lincoln Laboratory allowed graduate student Ivan Sutherland's Sketchpad and other early computer graphics projects to consume extremely expensive TX-2 cycles. Whether a coincidence or not, HCI thriving during downturns in interest in AI would be a recurring pattern.

The response to Sputnik soon revived AI. J. C. R. Licklider, as director of ARPA's Information Processing Techniques Office from 1962 to 1964, provided extensive support for computer science in general and AI in particular. MIT's Project MAC (originally Mathematics and Computation) was founded in 1963, focused on time-sharing and AI. It initially received US$13 million per year, rising to $24 million by 1969. ARPA sponsored the AI Laboratory at SRI, AI research at CMU, and Nicholas Negroponte's Architecture Machine Group at MIT. A dramatic early achievement, SRI's Shakey the Robot, was featured in articles in *Life* magazine and *National Geographic* in 1970.[32] Given a simple but non-trivial task, Shakey went to a specified location, scanned and reasoned about the surroundings, and moved objects as needed to accomplish the goal. (Videos of Shakey are on the SRI website and elsewhere.[33])

In 1970, Negroponte outlined a case for machine intelligence: "Why ask a machine to learn, to understand, to associate courses with goals, to be self-improving, to be ethical—in short, to be intelligent?" He noted common reservations: "People generally distrust the concept of machines that approach (and thus why not pass?) our own human intelligence." And he identified a key problem: "Any design procedure, set of rules, or truism is tenuous, if not subversive, when used out of context or regardless of context." This insight that it is risky to apply algorithms without understanding the situation at hand led Negroponte to a false inference: "It follows that a mechanism must recognize and understand the context before carrying out an operation."

---

[32]   Darrach, 1970; White, 1970.

[33]   http://www.ai.sri.com/shakey/. Nice accounts are found by entering "Shakey robot SRI video" in a browser.

An alternative is that the mechanism is guided by humans who understand the context: Licklider's human-machine symbiosis. Overlooking this, Negroponte sought funding for an ambitious AI research program:

> Therefore, a machine must be able to discern changes in meaning brought about by changes in context, hence, be intelligent. And to do this, it must have a sophisticated set of sensors, effectors, and processors to view the real world directly and indirectly... A paradigm for fruitful conversations must be machines that can speak and respond to a natural language... But, the tete-à-tete (sic) must be even more direct and fluid; it is gestures, smiles, and frowns that turn a conversation into a dialogue... Hand-waving often carries as much meaning as text. Manner carries cultural information: the Arabs use their noses, the Japanese nod their heads...

> Imagine a machine that can follow your design methodology, and at the same time discern and assimilate your conversational idiosyncrasies. This same machine, after observing your behavior, could build a predictive model of your conversational performance. Such a machine could then reinforce the dialogue by using the predictive model to respond to you in a manner that is in rhythm with your personal behavior and conversational idiosyncrasies... The dialogue would be so intimate—even exclusive—that only mutual persuasion and compromise would bring about ideas, ideas unrealizable by either conversant alone. No doubt, in such a symbiosis it would not be solely the human designer who would decide when the machine is relevant.

Negroponte's MIT colleague Minsky went further, as reported in a *Life* magazine article by a renowned reporter:

> In from three to eight years we will have a machine with the general intelligence of an average human being. I mean a machine that will be able to read Shakespeare, grease a car, play office politics, tell a joke, and have a fight. At that point, the machine will begin to educate itself with fantastic speed. In a few months, it will be at genius level and a few months after that its powers will be incalculable (Darrach, 1970).

Other AI researchers told Darrach that Minsky's timetable was ambitious:

> 'Give us 15 years,' was a common remark—but all agreed that there would be such a machine and that it would precipitate the third Industrial Revolution, wipe out war and poverty and roll up centuries of growth in science, education and the arts.

Darrach ended by quoting Ross Quillian:

I hope that man and these ultimate machines will be able to collaborate without conflict. But if they can't, we may be forced to choose sides. And if it comes to choice, I know what mine will be. My loyalties go to intelligent life, no matter in what medium it may arise.

Responding to the collective sense of urgency, ARPA initiated major programs in speech recognition and natural language understanding in 1971.

Minsky later grumbled that he was misquoted. However, such predictions were everywhere. In 1960, Nobel laureate and AI pioneer Herb Simon had written, "Machines will be capable, within twenty years, of doing any work that a man can do." The same year, an Air Force study concluded that by 1980, machines would be intelligent enough to handle military problems alone. In 1963, John McCarthy obtained ARPA funding to produce a "fully intelligent machine within a decade."[34] In 1965, Oxford mathematician I. J. Good wrote, "the survival of man depends on the early construction of an ultra-intelligent machine," and "the intelligence of man would be left far behind… It is more probable than not that, within the twentieth century, an ultra-intelligent machine will be built and that it will be the last invention that man need make." Indeed, if Minsky had not made such predictions in 1970, he might have had difficulty getting funded.

It is important to understand the anxieties of the time, and also the consequences of such claims. The world had barely avoided a devastating thermonuclear war during the Cuban missile crisis of 1962. Leaders seemed powerless to defuse the Cold War. As Good indicated, machines could save us! Humanity being saved by ultra-intelligence was also a theme in science fiction of the era. The possibility of a machine savior may have reassured some. Competition with the USSR also was a factor; Minsky estimated they were only three years behind us in AI, and a USSR chess-playing machine defeated a U.S. competitor designed by Stanford and MIT in a 1968 contest. The consequences for HCI were not good. Funds and good students gravitated to AI. An ultra-intelligent machine would be able to clean up all of the world's user interfaces, so why should anyone focus on such trivialities?

Ironically, central to funding the AI research was a psychologist who was not wholly convinced. In 1960, citing the Air Force study and his colleagues' forecasts, Licklider noted that until 1980, human efforts would be useful: "That would leave, say, 5 years to develop human-computer symbiosis and 15 years to use it. The 15 may be 10 or 500, but those years should be intellectually the most creative and exciting in the history of mankind."[35] Ten to five hundred years represents breathtaking uncertainty. Recipients of Licklider's AI funding were on the optimistic end of this spectrum. Speech and language recognition, which Licklider considered integral to achieving symbiosis, were well-funded.

---

[34] Moravec, 1988.
[35] Licklider's first five years correspond to what we call HCI. It has taken much longer to reach symbiosis. Licklider no doubt knew that it would, but he had decided not to contest openly his colleague's forecast that ultra-intelligence would arrive within 20 years.

Five years later, disappointed with the lack of progress, ARPA cut off AI support. A similar story unfolded in Great Britain. Through the 1960s, AI research expanded, spearheaded by Turing's former colleague Donald Michie. In 1973, the Lighthill report, commissioned by the Science and Engineering Research Council, was generally pessimistic about AI scaling up to address real-world problems. Government funding was cut.

The next decade has been called an AI winter, a recurring season in which research funding is withheld due to disillusionment over unfulfilled promises. The bloom was again off the rose, but it would prove to be a hardy perennial.[36]

5.5   LIBRARY SCHOOLS EMBRACE INFORMATION SCIENCE

Work on information science and "human information behavior" in the 1960s and 1970s focused on scholarship and application in science and engineering.[37] With "big science" alive and well post-Sputnik, aligning with national priorities was a priority for many researchers.

The terms "information science," "information technology," and "information explosion" came into use in this period. In 1968, the ADI became the American Society for Information Science (ASIS). Two years later, the journal *American Documentation* became *Journal of the American Society for Information Science*. In 1978 the ACM Special Interest Group on Information Retrieval (SIGIR) was formed and launched the annual Information Storage and Retrieval conference (since 1982, "Information Retrieval"), modeled on a conference held seven years earlier. By 1980, schools at over a dozen universities had added "information" to their titles, many of them library schools in transition. In 1984, the American Library Association belatedly embraced the i-word by creating the Association for Library and Information Science Education (ALISE), which convenes an annual research conference.

The pioneering University of Pittsburgh and Georgia Tech programs flourished. The former's Graduate School of Library and Information Sciences created an information science Ph.D. program in 1970, declaring humans to be "the central factor in the development of an understanding of information phenomena."[38] The program balanced behavioral science (psychology, linguistics, communication) and technical grounding (automata theory, computer science). Three years later, Pittsburgh established the first information science department, which developed an international reputation. The emphasis shifted slowly from behavior to technology. The Georgia Tech School of Information Science expanded after receiving an NSF center grant in 1966. In 1970, it became a Ph.D.-granting school, rechristened "Information and Computer Science."

Terminal-based computing costs continued to decline. The ARPANET debuted in 1969. It supported email in 1971 and file-sharing in 1973, which spurred visions of a "network society"

---

[36] More about AI summers and AI winters is found in Grudin (2009) and Markoff (2015).
[37] Fidel, 2011.
[38] Aspray, 1999.

(Hiltz and Turoff, 1978). However, developing and deploying transformative information technology could be difficult. A prominent example was MIT's Project Intrex, the largest unclassified information research project of the time. From 1965 to 1972, the Ford and Carnegie Foundations, NSF, DARPA, and the American Newspaper Publishers Association invested over US$30 million to create a "library of the future." Online catalogs were to include up to 50 index fields per item, accessible on CRT displays, with the full text of books and articles converted to microfilm and read via television displays. None of this proved feasible.[39]

As an aside, the optimism of this era lacks the psychological insight and nuance of the novelist E. M. Forster, who in 1909 anticipated both AI and a networked society in his remarkable *The Machine Stops*. In the story, the world is run by a benign artificially intelligent machine, with various effects on the human beings, including how to cope as the machine begins to fail.

---

[39] Burke, 1998.

CHAPTER 6

Hardware Generations

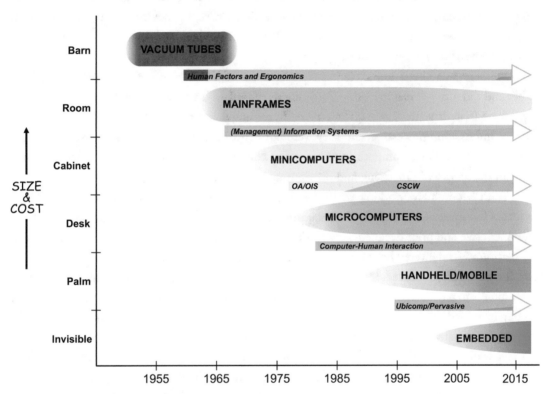

Figure 6.1: Hardware platforms and associated research fields.

6.1 THE PLATFORMS

The computer science research literature is over 60 years old. Papers that are more than 20 years old are almost impossible to understand, because of one misleading word: computer. We instinctively think of our current computer. Authors rarely describe a computer, because their immediate readers will understand. But readers a decade or two later won't understand. Even people who lived through the transition forget how radically computers changed in size, cost, and capability. As shown in Figure 6.1, every ten years a new platform appeared that was *two orders of magnitude smaller* than the previous. It was also *an order of magnitude less expensive*, yet *more capable* than the previous platform was when it appeared. Also keep in mind that authors who indicate what they paid for a computer usually overlook inflation: a dollar in the early days is equivalent to $10 today.

Processing power rose as more transistors were packed as integrated circuits on a chip. The integrated circuits that appeared in mainframes in 1964 comprised up to ten transistors. These computers filled a room. By 1971, large-scale integrated (LSI) circuits comprised 500 or more transistors, enabling cabinet-sized minicomputers. Often used in movies to represent powerful computers, they were not actually that powerful. The year 1980 heralded very large-scale integrated circuits, the heart of microcomputers.

Memory was also a key factor. The 300K bytes of TX-2 core memory with which Ivan Sutherland did groundbreaking work cost millions of dollars. Think about that in the context of a 32 gigabyte flash drive that costs a few dollars. Accordingly, computers were initially for computation, not information processing, and they were not networked. The situation is reversed in the era of data science and cloud computing: memory and bandwidth are so plentiful that most computation is in the service of processing and distributing information.

With Moore's law working year by year, why wasn't change continuous? It was—within a platform. The dynamics are nicely described in *The Innovator's Dilemma* (Christensen, 1997). As component size and price declined, new market possibilities opened up for smaller machines. Each new platform was delivered by companies that did not build the previous platform. Then, as a platform incrementally improved, it undercut the previous platform, delivering benefits great enough for customers to pay the costs of switching. Vacuum tube computers vanished when transistor-based computers arrived. Mainframes were a lucrative business for several manufacturers, but only IBM survived the onslaught of minicomputers, which in turn disappeared when organizations shifted to PCs and workstations. Today, PCs are losing ground to mobile devices. Just as organizations held onto one mainframe and bought many smaller computers, so a family might keep and occasionally replace a PC while acquiring a host of smartphones, e-readers, and other devices.

Handhelds such as the Apple Newton and the Palm Pilot appeared in the mid-90s, but uptake was slow because small devices require radically different input and display designs, novel user behavior (learning to enter text with a stylus or thumbs), and new approaches to electric power and data networking. The same is true with embedded computation and the Internet of Things: novelty greeting cards and magazine inserts appeared on schedule, but broad uptake may require harvested energy and complex networking solutions.

When computers of modest capability were few and extremely expensive, access was severely limited. The resource competition of that era is difficult to imagine today. Advanced work could only be done at a few research centers. Other computer science departments had difficulty recruiting, at one point offering faculty positions to graduates of good masters programs.

An important takeaway is that insight, design, *and timing* are all critical. Innovations were often held back until the processing power or memory to deliver them was affordable. Under-powered products appeared, failed, and were forgotten.

6.2   THE FIELDS

Figure 6.1 identifies research fields prominently associated with each hardware generation, including three of the four focused on in this book. Information was not closely tied to a platform and later thrived, with the shift from computation to information processing. The office automation (OA) or office information systems (OIS) field lived and died with minicomputers, as described in Chapter 7. Academic fields and professional organizations are less susceptible to market pressures than companies, but their fates can be linked: MIS reached its apogee in the mainframe era and CHI dominated when PCs ruled. The relatively slow emergence of handheld devices enabled several conferences and organizations to form around them and existing conferences to expand to embrace them, resulting in no single preferred locus.

Platform changes contributed to geographic shifts in innovation in the United States. Much of the early government-funded work was in the East, notably in Cambridge. Mainframes were produced primarily by IBM in the East and by several companies in Minnesota. The minicomputer companies were located around Boston. The focus began to move west in the 1960s. Much of the extraordinary cohort of Evans and Sutherland students at Utah went to California, including Alan Kay, William Newman, and Butler Lampson, who were key members of the team developing the Xerox Alto, and others who went on to found Pixar, Adobe, Silicon Graphics, and Netscape. In 1974, Sutherland himself moved to the California Institute of Technology. Engelbart at SRI and Xerox PARC were attractors for computer graphics researchers. When microcomputers arrived, Bay Area hobbyists picked them up and were ready for the PC. John Markoff's (2005) *What the Dormouse Said: How the 60s Counter-Culture Shaped the Personal Computer* explores the interaction of cultural and technological attractors in the Bay Area.

The history of trade shows also illustrates the influence of hardware platforms. In the 1970s and early 1980s, the National Computer Conference (NCC) was the premiere research conference and trade show in the U.S. It occupied a vast auditorium and earned millions of dollars for the American Federation of Information Processing Societies (AFIPS), the parent organization of ACM and IEEE that represented the U.S. in the International Federation for Information Processing (IFIP). NCC was fully aligned with the mainframe industry; other vendors were excluded. The minicomputer-based OA/OIS field formed a parallel research conference with an associated trade show, also hosted by AFIPS. The PC industry was drawn to the Computer Dealers' Exhibition (COMDEX), which began promoting itself as "the NCC for PCs." COMDEX had no need for a research track—by the 1980s, the commercial market had grown and specialized research conferences were proliferating. In 1988, the collapsing mainframe companies pulled out of NCC. AFIPS lost money and folded two years later. The OA/OIS enterprise folded soon after. COMDEX thrived for a decade but dwindled in the 2000s, when The Consumer Electronics Show (CES) became the preeminent U.S. trade show.

Europe, with fewer computer manufacturers, escaped much of this turmoil. Germany's CeBIT is the world's largest exhibition and has been held since 1970. After two decades of growth, Taipei's COMPUTEX became the second largest exhibition in 2003 when COMDEX ended.

CHAPTER   7

1980–1985: Discretionary Use Comes into Focus

In 1980, HF&E and IS were focused on the down-to-earth business of making efficient use of expensive mainframes. The beginning of a major shift went almost unnoticed. Less expensive but highly capable minicomputers based on LSI technology were competing for the low end of the mainframe market. Home computers based on microprocessors were gaining traction; the VisiCalc spreadsheet introduced for the Apple II in 1979 showed the potential for business. A third simple programming language, BASIC, enabled more people to write programs. (FORTRAN was for science; COBOL was for business; the BA in BASIC stood for Beginner's All-Purpose.) Students and hobbyists were drawn to these computers, creating a population of hands-on discretionary users. Experiments with online library catalogs and electronic journals had begun.

Then, between 1981 and 1984, a flood of innovative, powerful microcomputers were released: Xerox Star, IBM PC, Apple Lisa, Lisp machines from Symbolics and Lisp Machines, Inc. (LMI), workstations from Sun Microsystems and Silicon Graphics (SGI), and the Apple Macintosh. All supported interactive, discretionary use.

Relatively low-cost computers created markets for shrinkwrap software that targeted, for the first time, non-technical hands-on users who would get little or no formal training. It had taken 20 years, but the early visions were being realized! Non-programmers were choosing to use computers to do their work. The psychology of discretionary users intrigued two groups: (i) experimental psychologists who liked to use computers, and (ii) computer and telecommunications companies that desired to sell to discretionary users. Not surprisingly, the latter began hiring a lot of the former.

Personal and small business use of computers was held back by (i) the high cost of displays, which led to the awkward solution of hooking computers up to televisions, and (ii) printers, which were expensive and/or noisy and required maintenance. Less expensive dot matrix printers did not produce typewriter-quality output. In the early 1980s, display costs came down and resolution increased. Starting in 1984, HP's inkjet printers brought the cost of letter-quality printing, with what you see on the screen being what you get on paper (WYSIWYG, it was called), within reach of consumers. HP and Apple laser printers, appearing in 1984 and 1985, respectively, supported bitmap graphics and were affordable by small businesses. These printers were also quieter.

A change that arguably most strongly impacted the psychological perspective on technology and choice in the U.S. occurred on January 1, 1984, when AT&T's breakup into competing companies took effect. With more employees and more customers than any other U.S. company,

AT&T had been a monopoly: neither customers nor employees had much discretion in technology use. Accordingly, AT&T and its Bell Laboratories research division had employed human factors to improve training and operational efficiency. Suddenly, customers of AT&T and the new regional operating companies had choices. Until it ceased being a monopoly, AT&T was barred from the computer business, but it had been preparing. In 1985, it launched the Unix PC. AT&T lacked experience designing for discretionary use, which is why you haven't heard of the Unix PC. The HCI focus of telecommunications companies had to broaden.[40]

Discretion in Computer Use

Technology use lies on a continuum, from the assembly line nightmare of *Modern Times* to utopian visions of completely empowered individuals. To use a technology or not to use it: sometimes we have a choice; other times we do not. At home, computer use is largely discretionary. We may have to wrestle with speech recognition routing systems on the phone. The workplace often lies in between: technologies are prescribed or proscribed, but we can ignore some injunctions or obtain exceptions, choose which features to use, and join with colleagues to press for changes.

For early computer builders, the work was more a calling than a job, but a staff was needed to carry out essential but less interesting tasks. For half of the computing era, most hands-on use was by computer operators, date entry clerks, and others who had a mandate to make use of every expensive minute. Over time, hardware innovation, more versatile software, and progress in understanding the tasks and psychology of users—and transferring that understanding to software developers—led to hands-on users who had some control over how they worked. Rising expectations played a role; people heard that software is flexible. Competition among vendors brought more emphasis on "user friendliness."

Discretion is not all-or-none. An airline reservation clerk must use a computer; a traveler booking a flight still has a choice but may get a fare discount for booking online. Many jobs and pastimes require some computer use, but people can resist or quit. A clerk or a systems administrator has less discretion than someone engaged in an online leisure activity. *These distinctions and the shift toward greater discretion are at the heart of the history of HCI.*

Although the inspirational writers of the 1960s envisioned discretionary use, it took time to get there. John Bennett (1979) predicted that discretionary use would lead to more emphasis on usability. The 1980 book *Human Interaction with Computers*, edited by Harold Smith and Thomas Green (1980), stood on the cusp. A chapter by Jens Rasmussen, "The Human as a Systems Component," covered the nondiscretionary perspective. One-third of the book covered research on programming. The remainder addressed "non-specialist people," discretionary users who are not computer-savvy. Smith and Green wrote, "It's not

---

[40] Israelski and Lund, 2003.

enough just to establish what computer systems can and cannot do; we need to spend just as much effort establishing what people can *and want to do*." The italics are in the original, indicating that this was a new imperative.

A decade later, Liam Bannon (1991) noted broader implications of the shift taking place "from human factors to human actors." But the trajectory is not always toward choice. Discretion can be curtailed; for example, word processor and email use became job requirements, not an optional alternative to typewriters and phones. Mandatory use remains important. Although greater competition, customization, and specialization lead to more choice, how it is exercised varies over time and across contexts.

Priorities differ in mandatory versus discretionary contexts. With heavily used systems, such as those handling taxes or census collection, efficiency is prized and there is a willingness to invest in training. Error reduction is prioritized in mission-critical systems, such as air traffic control. In contrast, for discretionary applications, first-time and occasional use scenarios are important. There is less reliance on training, less concern for split-second efficiency, and often more forgiveness for occasional errors.

HFES and IS, which arose when computer time was expensive, focused on mandatory use. Gilbreth worked on labor-saving household products and there are product development and organizational systems technical groups in HFES, but the human factors field was formed by aviation psychologists and retains to this day a large-system, non-discretionary emphasis. Minicomputers and personal computers reduced the cost of computer time and expanded the reach of discretionary use; the different priorities gave rise to different methods. Discretion is only one factor in understanding HCI history, but analysis of its role casts light on how efforts differed across HCI disciplines and why they did not interact more.

7.1   MINICOMPUTERS AND OFFICE AUTOMATION

Cabinet-sized minicomputers that could support several people simultaneously had arrived in the mid-1960s. By the late 1970s, super-minis such as the VAX 11/780 supported integrated suites of productivity tools. In 1980, the leading minicomputer companies, Digital Equipment Corporation, Data General, and Wang Laboratories, were a growing presence outside Boston.

Because of their smaller size, minicomputers became the first embedded systems, and were used to control medical devices such as CT scanners, manufacturing process control, and military systems. Today, most of the computers we use are embedded in cars, microwaves, and other devices. We interact with them, but we see the task, not the computer. However, minicomputers also had a very visible impact.

A minicomputer could handle personal productivity tools or a database of moderate size. Users sat at terminals. A terminal could be "dumb," passing each keystroke to the central processor, or it could contain a processor that supported a user entering a screenfull of data that was then sent on command as a batch to a nearby central processor. Minis provided a small group (or "office") with file sharing, word processing, spreadsheets, email, and managed output devices. They were marketed as "office systems," "office automation systems," or "office information systems."

The 1980 Stanford International Symposium on Office Automation had two papers by Douglas Engelbart.[41] It marked the emergence of a research field that remained influential for a decade, and then faded away. Also in 1980, ACM formed the Special Interest Group on Office Automation (SIGOA), and AFIPS held the first of seven annual office automation conferences and product exhibitions. In 1982, SIGOA initiated the biennial Conference on Office Information Systems (COIS) and the first issue of the journal *Office: Technology and People* appeared, followed in 1983 by *ACM Transactions on Office Information Systems* (*TOOIS*).

You might be wondering, "What is all this with offices?" Minicomputers reduced the price of a computer from an enterprise decision to one within the budget of a small work-group: an office. (The attentive reader will anticipate: The personal computer era was approaching! It was next!) The field of office information systems (OIS), focused on minicomputer use, was positioned alongside management information systems (MIS), which focused on mainframes. The research scope was reflected in the charter of *TOOIS*: database theory, AI, behavioral studies, organizational theory, and communications. Database researchers could afford to buy minis. Digital's PDP series was a favorite of AI researchers until Lisp machines arrived. Minis were familiar to behavioral researchers who used them to run and analyze experiments. Computer-mediated communication (CMC) was an intriguing new capability: networks were rare, but people at different terminals, often in different rooms, could exchange email or chat in real time. Minis were the interactive computers of choice for many organizations. Digital became the second largest computer company in the world. Wang Laboratories developed a hugely successful minicomputer-based word processor, and its founder Dr. An Wang became the fourth wealthiest American.

Researchers were discretionary users, but few office workers chose their tools. The term "automation" (in "office automation") was challenging and exciting to researchers, but it conjured up less pleasant images for office workers. And some researchers also preferred Engelbart's focus on augmentation.

Papers in the SIGOA newsletter, COIS, and TOOIS included database theory and technology, a modest stream of AI papers (the AI winter of the mid-1970s had not yet ended), decision support and CMC papers from the information systems community, and behavioral studies by researchers who later joined CHI. The SIGOA newsletter published short IS research papers. *TOOIS*

---

[41]   Landau et al., 1982.

favored more technical work and was also a major outlet for behavioral studies until the journal *Human-Computer Interaction* started in 1985.

Although OA/OIS research was eventually absorbed by other fields, it identified important emerging topics, including object-oriented databases, hypertext, computer-mediated communication, and collaboration support. OIS research also conceptually reflected aspects of the technical side of information science, notably information retrieval and language processing.

7.2   PCs AND COMMUNITY BULLETIN BOARDS

The general-purpose PC emerged in 1981 from what was almost a skunkworks project at IBM. Few anticipated its success, much less that it would eradicate the huge minicomputer market within a decade. PCs were initially difficult to network and came with little software, and printers were still too expensive for most consumers. But it was much less expensive than an office system—an employee could order one without a major requisition process and use it to manage spreadsheets.

Another major product release in 1981 was the Hayes Smartmodem, which greatly simplified connecting to a computer via a phone line. Together, these led to the proliferation of community bulletin board systems (BBSs) in the U.S., where a host computer could be reached toll-free within an area code. The model was the Berkeley Community Memory system that in the mid-70s ran on a mainframe that residents accessed from coin-operated terminals in a few locations around the city. With a PC and modem, people could participate from home.

Although not prominent in mainstream media or with academics using minicomputers, BBSs enabled consumers to develop computer skills and get familiar with asynchronous, distributed interaction in relatively non-threatening, local contexts. They were widely used through the 1980s and into the mid-90s, at which point there were an estimated 60,000 BBSs serving 17 million users.

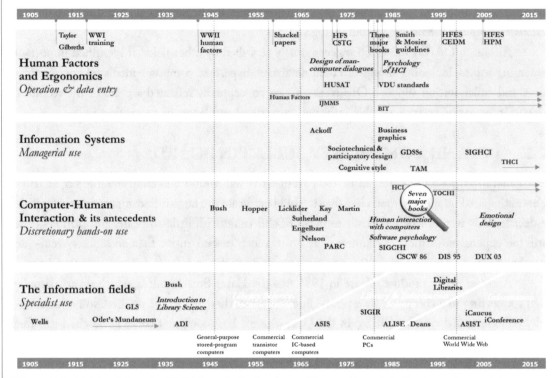

Figure 7.1: Fields with major HCI threads. Left edges of items align with the dates that articles or books were published, organizations or conference series initiated, and so on. Details are in the text.

7.3   THE FORMATION OF ACM SIGCHI

Figure 7.1 identifies the research fields that have the most direct bearing on HCI. Human factors and information systems have distinct subgroups that focus on digital technology use. The relevant computer science research is concentrated in CHI, the interest group primarily concerned with discretionary hands-on computer use. Other CS influences—computer graphics, AI, and OIS—are discussed in the text but are not broken out in Figure 7.1. The fourth field, information, began as support for specialists but has broadened as personal information management expanded, especially since 2000.

In the 1970s, it dawned on some people that software could have value. Hardware was so profitable that companies sometimes did not bother to track the cost of developing software and generally included their software applications with a hardware purchase, a practice known as "bundling." IBM's dominant position in mainframes led to litigation that forced it to unbundle software in 1969, opening that market to competition and leading IBM to make software a product focus.

The hobbyists attracted to relatively inexpensive microcomputers such as the Apple II and the Commodore PET created demand for applications that ran on them. As recounted in Stephen Levy's *Hackers* (1984), the culture of freely sharing software was challenged in 1976 by Bill Gates's "Open Letter to Hobbyists"[42] protesting the unauthorized copying of his BASIC software for Altair 8800 microcomputers. In 1980, as IBM prepared to launch the PC, a groundswell of attention to the behavior of potential software users was building.

Some experimental psychologists were using minicomputers to run and analyze experiments. Cognitive psychologists studying human memory and human information processing were naturally introspected on their thought processes as they programmed and used computers. Computer and telecommunications company interest in their skills coincided with a weakening academic cognitive psychology job market. At IBM, John Gould, who had published human factors research since the late 1960s, initiated empirical studies of programming and studies of software design and use, working with a growing team of psychologists. Other psychologists who led recently formed HCI groups in 1980 included Stu Card, who with Tom Moran had moved from CMU to Xerox PARC; Phil Barnard at the Medical Research Council Applied Psychology Unit in (MRC APU) Cambridge, England (partly funded by IBM and British Telecom); Tom Landauer at Bell Laboratories; John Whiteside at Digital Equipment Corp; and Donald Norman at the University of California, San Diego.

From one perspective, CHI was formed by psychologists who saw an opportunity to shape a better future. From another, it was formed by managers in computer and telecommunications companies who saw that digital technology would soon be in the hands of millions of technically unsophisticated users with unknown interaction needs. Invention? Or incremental improvement based on empirical observations? Competing views of CHI's mission were present from the outset.

Communications of the ACM established a "human aspects of computing" focus in 1980, edited by Whiteside's mentor Henry Ledgard. The next year, Tom Moran of Xerox PARC edited a special issue of *Computing Surveys* on "The Psychology of the Computer User." Also in 1981, the ACM Special Interest Group on Social and Behavioral Science Computing (SIGSOC) extended its workshop to cover interactive software design and use. A 1982 conference in Gaithersburg, Maryland, on "Human Factors in Computing Systems" was unexpectedly well attended. SIGSOC shifted its focus to computer-human interaction and changed its name to SIGCHI.[43]

In 1983, the first CHI conference attracted more than 1,000 people. Half of the 58 papers were from the seven organizations mentioned above (IBM, Xerox PARC, CMU, MRC APU, Bell Labs, Digital, and UCSD). Cognitive psychologists working in industry dominated the program, but HFS co-sponsored the conference with SIGCHI and was represented by program chair Richard Pew, committee members Sid Smith, H. Rudy Ramsay, and Paul Green, and several paper

---

[42] http://www.digibarn.com/collections/newsletters/homebrew/V2_01/index.html.
[43] Borman, 1996.

presenters. Brian Shackel and HFS president Robert Williges gave tutorials the first day. The alliance continued in 1984, when HF&E and CHI researchers attended the first IFIP International Conference on Human-Computer Interaction (INTERACT) in London, chaired by Shackel.

The first profession to enjoy discretionary hands-on computer use was computer programming. It did not start that way. Many early programmers had begun as students or hobbyists who chose to use computers, but they typically wrote on paper coding sheets and punched program instructions on cards that operators loaded into a computer. Eventually, when they could, most programmers switched to using text editors or other tools on interactive terminals, PCs, and small minicomputers. Early CHI papers by Ruven Brooks, Bill Curtis, Thomas Green, Ben Shneiderman, and others continued the psychology-of-programming research thread. Shneiderman formed the influential HCI Laboratory (HCIL) at Maryland in 1983. IBM's Watson Research Center contributed, as noted by John Thomas:

> One of the main themes of the early work was basically that we in IBM were afraid that the market for computing would be limited by the number of people who could program complex systems, so we wanted to find ways for "nonprogrammers" to be able, essentially, to program.[44]

The prevalence of experimental psychologists studying text editing was captured by Thomas Green at INTERACT'84: "Text editors are the white rats of HCI." As personal computing spread and researchers increasingly focused their attention on "non-programmers," studies of programming gradually disappeared from HCI conferences. For people who can choose whether or not to use a system, the initial experience is significant, so first-time use became important. New users were also a natural focus given that each year, more people took up computer use for the first time than in the previous year.

Of course, routinized expert use was still widespread. Databases were used by airlines, banks, government agencies, and other organizations. In these settings, hands-on activity was rarely discretionary. Managers oversaw software development and data analysis, leaving data entry and information retrieval to people hired for those jobs. To improve skilled performance required a human factors approach. CHI studies of database use were few—I counted three over the first decade, all focused on novice or "casual" (occasional) use.

CHI's emphasis differed from the HCI arising in Europe. Few European companies produced mass-market software. European HCI research focused on in-house development and use, as reflected in articles published in *Behaviour and Information Technology*, a journal launched in 1982 by Tom Stewart in London. In the perceptive essay cited earlier, Liam Bannon (1991) urged that more attention be paid to discretionary use, but he also criticized CHI's emphasis on initial experiences, reflecting a European perspective. At Loughborough University, HUSAT focused on

---

[44] John Thomas, personal communication (October, 2003).

job design (the division of labor between people and systems) and collaborated with the Institute for Consumer Ergonomics, particularly on product safety. In 1984, Loughborough initiated an HCI graduate program drawing on human factors, industrial engineering, and computer science.

Many CHI researchers had not read the early visionaries, even as they labored to realize some of the visions. The 633 references in the 58 CHI'83 papers included work of well-known cognitive scientists, but Vannevar Bush, Ivan Sutherland, and Nelson were not cited. A few years later, when computer scientists and engineers (primarily from computer graphics) joined CHI, the psychologists learned about the pioneers who shared their interest in discretionary use. By appropriating this history, CHI acquired a sense of continuity that bestowed legitimacy on a young enterprise seeking to establish itself academically and professionally.

7.4   CHI AND HUMAN FACTORS DIVERGE

"Hard science, in the form of engineering, drives out soft science, in the form of human factors."

— Newell and Card (1985)

A heated war raged in psychology through the 1960s and 1970s and into the 1980s. The behaviorists, with a long history culminating in the work of B. F. Skinner, focused on measurable behavior and avoided theorizing about internal processes and structures. Cognitive psychology drew support from the computer as a model—computers unquestionably had internal memory and processes that were critical to understanding their behavior (output). Human factors leaders trained in empirical behavioral psychology were adamantly opposed to cognitive theorizing about human behavior: "They are pursuing unobservable will o' the wisps!"[45] The cognitive psychologists accused the behaviorists of ignoring memory, problem-solving, and other key phenomena. The Newell and Card quotation above reflects the impasse: human factors engineers considered themselves "hard" scientists; one ridiculed the "thinking-aloud" protocols favored by some cognitive psychologists as "people talking about their feelings."

Card et al. (1980a,b) introduced a "keystroke-level model for user performance time with interactive systems." This was followed by the cognitive model GOMS—goals, operators, methods, and selection rules—in their landmark 1983 book, *The Psychology of Human–Computer Interaction*. Although highly respected by CHI cognitive psychologists, these models did not address discretionary, novice use. They modeled the repetitive expert use studied in human factors. GOMS was explicitly positioned to counter the stimulus-response bias of human factors research:

Human–factors specialists, ergonomists, and human engineers will find that we have synthesized ideas from modern cognitive psychology and artificial intelligence with the old

---

[45]   A leading human factors engineer, quoted by Clayton Lewis, personal communication (August, 2016).

methods of task analysis… The user is not an operator. He does not operate the computer, he communicates with it…

Newell and Card (1985) noted that human factors had a role in design, but continued:

Classical human factors… has all the earmarks of second-class status. (Our approach) avoids continuation of the classical human–factors role (by transforming) the psychology of the interface into a hard science.

In 2004, Card noted in an email discussion:

Human Factors was the discipline we were trying to improve… I personally changed the call [for CHI'86 participation], so as to emphasize computer science and reduce the emphasis on cognitive science, because I was afraid that it would just become human factors again.

Human performance modeling drew a modest but fervent CHI following. Their goals differed from those of most other researchers and many practitioners as well. "The central idea behind the model is that the time for an expert to do a task on an interactive system is determined by the time it takes to do the keystrokes," wrote Card et al. (1980b). Although subsequently extended to a range of cognitive processes, the modeling was used to design for nondiscretionary users, such as telephone operators engaged in repetitive tasks.[46] Its role in augmenting human intellect was not evident.

"Human Factors in Computing Systems" remains the CHI conference subtitle, but CHI and human factors moved apart without ever having been highly integrated. The cognitive psychologists had turned to HCI after earning their degrees and most were unfamiliar with the human factors literature. HFS did not cosponsor CHI after 1985, and their researchers disappeared from CHI program committees. Most CHI researchers who had published in the human factors literature shifted to the CHI conference, *Communications of the ACM*, and the Erlbaum journal *Human–Computer Interaction*, edited by Thomas Moran. Erlbaum primarily published experimental psychology books and journals.

The shift was reflected at IBM Watson. John Gould trained in human factors and would become president of the Human Factors Society. Clayton Lewis was a cognitive psychologist. Together they authored a CHI' 83 paper that best captured the CHI focus on user-centered, iterative design based on building and testing prototypes. Cognitive scientists at IBM helped shape CHI. In 1984, IBM's Human Factors Group dissolved and a User Interface Institute emerged.

CHI researchers and developers, wanting to identify with "hard science" and engineering, adopted the terms "cognitive engineering" and "usability engineering." In the first paper presented at CHI'83, "Design Principles for Human–Computer Interfaces," Donald Norman applied engi-

[46] Gray et al., 1990.

neering techniques to discretionary use, creating "user satisfaction functions" based on technical parameters. These functions did not hold up long—what was wonderful yesterday is not as satisfying today—but it would be years before CHI loosened its identification with engineering enough to welcome disciplines such as ethnography and design.

7.5   WORKSTATIONS AND ANOTHER AI SUMMER

High-end workstations from Apollo, Sun Microsystems, and Silicon Graphics arrived between 1981 and 1984. Graphics researchers no longer had to flock to heavily financed laboratories; most were at MIT and Utah in the 1960s, and MIT, PARC, and New York Institute of Technology in the 1970s. Workstations were priced beyond the reach of the mass market, so graphics work on photorealism and animation, which required the processing power of these machines, did not directly influence HCI.

The Xerox Star (formally named Office Workstation), Apple Lisa, and other commercial GUIs appeared, but when the first CHI conference convened in December 1983, none were commercial successes. They cost too much or ran on processors that were too weak to exploit interactive graphics effectively.

In 1981, Symbolics and LMI introduced workstations that were optimized for the Lisp programming language favored by AI researchers. The timing was fortuitous. In October of that year, the Next Generation Technology conference was held in the National Chamber of Commerce auditorium in Tokyo. In 1982, the Japanese government established the Institute for New Generation Computer Technology (ICOT) and launched a ten-year Fifth Generation project. AI researchers in Europe and the United States sounded the alarm. Donald Michie of Edinburgh called it a threat to Western computer technology. In 1983, Raj Reddy of CMU and Victor Zue of MIT criticized ARPA's mid-1970s abandonment of speech processing research. Also in 1983, Stanford's Ed Feigenbaum and Pamela McCorduck used the term "singularity" to describe the achievement of ultra-intelligence, writing:

> The Japanese are planning the miracle product… They're going to give the world the next generation—the Fifth Generation—of computers, and those machines are going to be intelligent… We stand, however, before a singularity, an event so unprecedented that predictions are almost silly… Who can say how universal access to machine–intelligence—faster, deeper, better than human intelligence—will change science, economics, and warfare, and the whole intellectual and sociological development of mankind?

Parallel distributed processing (also called "neural network") models seized the attention of researchers and the media. A conceptual advance over early AI work on perceptrons, neural networks were used to model signal detection, motor control, semantic processing, and other phenomena. Minicomputers and workstations were powerful enough to support simulation experiments. Production systems, another computer-intensive AI modeling approach with a psychological foundation, was developed at CMU.

An AI gold rush ensued. As with actual gold rushes, most of the money was made by those who outfitted and provisioned the prospectors, with generous government funding flowing once again to researchers. The European ESPRIT and UK Alvey programs invested over US$200 million per year starting in 1984. In the U.S., funding for DARPA's AI-focused Strategic Computing Initiative began in 1983 and rose to almost $400 million for 1988. AI investments by 150 U.S. corporations were estimated to total $2 billion in 1985, with 400 academics and 800 corporate employees working on natural language processing alone. These figures omit classified intelligence and military projects.[47]

The unfulfilled promises of the past led to changes this time around. General problem-solving was emphasized less, domain-specific problem-solving more. The term AI was used less, in favor of intelligent knowledge-based systems, knowledge engineering, expert systems, machine learning, language understanding, image understanding, neural networks, and robotics.

U.S. antitrust laws were relaxed in 1984 to permit 20 U.S. technology companies to form a research consortium, the Microelectronics and Computer Technology Corporation (MCC), an explicit response to the Japanese Fifth Generation project. MCC embraced AI, reportedly becoming the leading purchaser of Symbolics and LMI workstations. Within its broad scope were a large natural language understanding effort and work on intelligent advising. The centerpiece was Doug Lenat's CYC (as in "encyclopedia") project, an ambitious effort to build a common-sense knowledge base that other programs could exploit. Lenat predicted in 1984 that by 1994 CYC would be intelligent enough to educate itself. In 1989, he reported that CYC was on schedule and would soon "spark a vastly greater renaissance in [machine learning]."

In 1987, the "Knowledge Navigator" video (available on YouTube) would be released by Apple CEO John Sculley. It depicts a fluent, English-speaking, bowtie-wearing digital assistant carrying out a range of intelligent retrieval and modeling tasks in response to requests. Marketing materials described its features as under development, although close inspection of the video placed the fictional scenario in 2010. The Desktop videoconferencing (between humans) shown in Knowledge Navigator did materialize by 2010, but the AI capabilities did not.

A mismatch between the billions of dollars invested annually in speech and language processing and the revenue produced was documented in comprehensive Ovum reports in 1985 and 1991.[48] In 1985, "revenue" (mostly income from grants and investor capital, not sales) reached US$75 million. Commercial natural language understanding interfaces to databases were developed

---

[47]   Alvey funding: Oakley, 1990; **Strategic Computing Initiative:** Norberg and O'Neill, 1996; corporate investments: Kao, 1998, and Johnson, 1985.

[48]   Johnson, 1985; Engelien and McBryde, 1991.

and marketed: In 1983, Artificial Intelligence Corporation's Intellect became the first third-party application marketed by IBM for their mainframes. In 1984, Clout was the first AI product for microcomputers. With no consumer market materializing, the few sales were to government agencies and large companies where use was by trained experts.

Many AI systems focused on knowledge engineering: representing and reasoning about knowledge *obtained from experts*. European funding directives explicitly dictated that work cover both technology and behavior. This led to two challenges. (i) Eliciting knowledge from experts is difficult. (ii) Few AI professionals were interested in interaction details, as evidenced by their tolerance for the painful weeks required to master the badly designed command languages of powerful tools such as EMACS and Unix. Frustrated AI researchers collaborated with HF&E researchers, who shared the expert, non-discretionary focus of much of this work. The journal *IJMMS* became a major outlet for HF&E and AI research in the 1980s.

In contrast, early CHI conferences had a few papers on speech and language, cognitive modeling, knowledge-based help, and knowledge elicitation, but AI was not a significant focus. Despite meager results, hope springs eternal. Ovum concluded its 1985 review by predicting a 1,000% increase in speech and language revenue in five years: US$750 million by 1990 and $2.75 billion by 1995. Its 1991 review reported that revenue had increased by less than 20%, not reaching $90 million. Undaunted, Ovum forecast a 500% increase to $490 million by 1995 and $3.6 billion by 2000. But soon another winter set in. AI Corporation and MCC disappeared.

CHAPTER 8

1985–1995:
Graphical User Interfaces Succeed

"There will never be a mouse at the Ford Motor Company."

— High-level acquisition manager, 1985

The commercial success of the graphical user interface, when it finally arrived, was a disruptive revolution within HCI. As with previous major shifts—to stored programs, and to interaction based on commands, full-screen forms, and full-screen menus—some people were affected before others. GUIs were especially attractive to new users and people who were not chained to a computer but used them occasionally. GUI success in late 1985 immediately transformed CHI research, but only after Windows 3.0 succeeded in 1990 did GUIs influence the government agencies and business organizations that guided HF&E and IS. By then, GUI design was better understood. The early 1990s also saw the maturation of local area networking and the internet, which produced a second transformation: computer-mediated communication and information sharing.

8.1   CHI EMBRACES COMPUTER SCIENCE

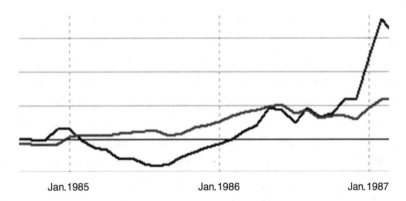

Figure 8.1: Apple stock price (black) and overall market (red). In mid-1985, the Macintosh was failing and Steve Jobs was forced out.

Apple launched the Macintosh with a 1984 Super Bowl commercial that is considered one of the best ads ever produced, easily found on YouTube. It envisioned the Mac supplanting IBM in the office. That did not happen. Apple had bet on the Mac and with sales languishing, in mid-1985, the company was in trouble (Figure 8.1). Steve Jobs was forced out. Months later, the "Fat Mac" was released. It had four times as much random-access memory (RAM), enough to run applications developed for the Mac: Aldus PageMaker, Adobe Postscript, the Apple LaserWriter, and Microsoft's Excel and Word. The more powerful Mac Plus arrived in January 1986, and Apple's recovery gained momentum. Rescued by hardware and software advances and declining prices, the Mac succeeded where the commercial GUIs that preceded it, including the Apple Lisa, had not. It was popular with consumers and became the platform for desktop publishing.

Following Jobs' departure, Larry Tesler built a group at Apple that focused on user studies and interface design to refine the key GUI concepts that he and others had pioneered at PARC. The Apple Human Interface Guidelines, published in 1987, sought to bring internal and external application developers to a common look and feel. Apple was a dominant force through the decade. Apple researchers participated actively in CHI conferences until Jobs returned in 1997.

Within CHI, GUIs were initially controversial. They had disadvantages. An extra level of interface code increased development complexity and decreased reliability. They consumed processor cycles and distanced users from the underlying system that, many believed, experienced users would eventually want to master. Carroll and Mazur (1986) showed that GUIs confused and created problems for people familiar with existing interfaces. An influential essay by Hutchins et al. (1986) on direct manipulation interfaces concluded that "It is too early to tell" how GUIs would fare. GUIs might well prove useful for novices, they wrote, but "we would not be surprised if experts are *slower* with Direct manipulation systems than with command language systems" (italics in original). However, GUIs were here to stay. As noted in the previous chapter, most pre-CHI HCI research had focused on expert use, but first-time use was critical in the rapidly expanding consumer market. CHI quickly came on board with studies that addressed GUI limitations, largely abandoning research into command names, text editing, and the psychology of programming.

As topics such as "user interface management systems" became prominent, psychology gave way to computer science as the driving force in interaction design. Early researchers had worked one formal experiment at a time toward a comprehensive psychological framework.[49] Such a theoretical framework was conceivable when interaction was limited to commands and forms, but it could not be scaled to design spaces that included color, sound, animation, and an endless variety of icons, menu designs, window arrangements, and input devices. The new mission: to identify the most pressing problems and find satisfactory rather than optimal solutions. Rigorous experimentation, a skill of cognitive psychologists, gave way to quicker, less precise assessment methods, a controversial shift promoted by Jakob Nielsen.[50] These methods were quickly adopted by practitioners, but researchers sought general "statistically significant" effects that required larger sample sizes.

---

[49] E.g., Newell and Card, 1985; Carroll and Campbell, 1986; Long, 1989; Barnard, 1991.
[50] Nielsen, 1989; Nielsen and Molich, 1990.

The significant "practitioner" community soon extended beyond those working on new products to include consultants and employees who worked for the companies who bought and were customizing the software. They faced similar issues and could apply the same techniques, but they were constrained by the purchased product and often supported its use beyond the initial contact. They often shared best practices in local SIGCHI chapters that sprang up and might not attend the national conference even when it met in their city.

Software engineering expertise was required to explore this dynamically evolving design space. In the late 1980s, the CHI community enjoyed an influx of computer scientists focused on interactive graphics, software engineers interested in interaction, and a few AI researchers working on speech recognition, language understanding, and expert systems. Many computer science departments added HCI to their curricula. In 1994, ACM launched *Transactions on Computer-Human Interaction* (*TOCHI*), which stressed short, technology-centered studies as *Human Computer Interaction* remained an outlet for behavioral HCI research. Early PCs and Macs were not easily networked, but as local area networks spread, CHI's focus expanded to include collaboration support, discussed later in this chapter.

The years 1986 to 1988 brought a remarkable intellectual development: seven books appeared that were widely read by researchers and practitioners. This created a shared understanding across the field that endured. *User Centered System Design: New Perspectives on Human-computer Interaction*, edited by Donald Norman and Steven Draper (1985), was comprised of ground-breaking essays on concepts and methods that retained a long shelf-life for a rapidly-moving field. *Readings in Human-Computer Interaction: A Multi-Disciplinary Approach* was a large set of key papers collected by Ronald Baecker and William Buxton (1987), who also wrote section and chapter introductions containing overviews, supplementary information, and references. The first *Handbook of Human-Computer Interaction*, edited by Martin Helander (1988), comprised 52 original chapters by leading researchers and a handful of practitioners. Don Norman's (1988) *Psychology of Everyday Things*, later reissued as *Design of Everyday Things*, is probably the most widely-read HCI book. Ben Shneiderman's (1986) *Designing the User Interface: Strategies for Effective Human–Computer Interaction* was the first textbook. In *Computer Supported Cooperative Work: A Book of Readings*, Irene Greif (1985) collected papers ranging from early Doug Engelbart to recent work. Finally, Lucy Suchman's (1987) *Plans and Situated Action* critiqued AI and promoted social science for understanding the contexts of technology use. Influential works preceded and followed these, but this burst of widely embraced books was unique.

One consequence of the market value of the "look and feel" of software was that it became the subject of litigation. Lotus sued Borland and other competitors in 1987; Apple sued Microsoft and Hewlett Packard in 1988; and Xerox sued Apple in 1989. Previously, interface concepts had been borrowed freely. Andy Hertzfeld of Apple recounted a meeting at which Steve Jobs confronted Bill Gates about Microsoft's graphical interface, shouting "You're ripping us off! I trusted you, and now you're stealing from us!" To which Gates replied, "Well, Steve, I think there's more than one way of looking at it. I think it's more like we both had this rich neighbor named Xerox and I broke into his house to steal the TV set and found out that you had already stolen it."[51]

Almost none of the lawsuits succeeded. However, just as the hacker community had been affected a decade earlier by the introduction of copyrighted software, this changed how HCI professionals viewed their roles. At CHI'89, Pamela Samuelson of the University of Pittsburgh Law School organized a session with prominent attorneys, "Protecting User Interfaces through Copyright: The Debate," and a panel discussion sequel at CHI'91. Some HCI researchers became (and remain) well-paid expert witnesses, but the community had mixed feelings.

8.2 HUMAN FACTORS & ERGONOMICS MAINTAINS A NONDISCRETIONARY USE FOCUS

After SIGCHI formed, HFS undertook a study to determine the effect on membership in its Computer Systems Technical Group. It found an unexpectedly small overlap.[52] The two organizations had different customers and methods, directly linked to the distinctions between discretionary and non-discretionary use. HF&E addressed military, aviation, and telecommunications industries and government use in general for census, tax, social security, health and welfare, power plant operation, air traffic control, space missions, military logistics, intelligence analysis, and so on. Government remained the largest customer of computing. Technology was assigned and training was provided. For repetitive tasks such as data entry, a very small efficiency gain in an individual transaction can yield huge benefits over time. This justified rigorous experimental human factors studies to make improvements that would go unnoticed by consumers and were of no interest to the fast-moving commercial software developers employing CHI researchers.

Government agencies also promoted the development of ergonomic standards. Standards enable system requirements to be defined for competitive contracts while staying at arms' length from potential bidders. They were regarded warily by competitive commercial software developers, who feared standards would constrain innovation. In 1986, human factors researchers Sid Smith and Jane Mosier (1986) published the last of a series of government-sponsored interface guidelines, 944 in all, organized into sections titled Data Entry, Data Display, Data Transmission, Data Protection, Sequence Control, and User Guidance. Smith and Mosier did not cover GUIs, but they recognized the implications of the Macintosh's recent success. Icons, pull-down and pop-up menus, mouse button assignments, sound, and animation would expand the design space far beyond the reach of an already cumbersome document. They foresaw that future contracts would specify predefined *interface styles and design processes*, rather than specific interface features.

---

[51] Isaacson, 2014.
[52] Richard Pew, personal communication (September 2004).

HF&E membership doubled in the 1980s. Much was peripheral to HCI, such as risk and safety—seat belt deployment and the Three Mile Island and Challenger shuttle disasters marked this decade—but the Computer Science Technical Group was active. Keyboard optimization for data entry was tractable, but many human factors tasks were affiliated with doomed AI efforts. DARPA's heavily funded Strategic Computing Initiative ended in 1993, having failed to develop an Autonomous Land Vehicle, a Pilot's Associate, and a Battle Management system.[53] Interfaces were sought for interactive technologies such as speech recognition, language understanding, and heads-up displays for pilots, drivers of autonomous vehicles, and officers working under stressful conditions. Unfortunately, the underlying functionality remained as elusive in the 1980s and 1990s as it was when Licklider at DARPA funded it in the 1960s and 1970s.

NSF, located in Washington, D.C., was heavily influenced by government initiatives. Its funding of HCI was largely directed to HF&E. Its Interactive Systems Program—subsequently renamed Human–Computer Interaction—was described as follows:

> The Interactive Systems Program considers scientific and engineering research oriented toward the enhancement of human-computer communications and interactions in all modalities. These modalities include speech/language, sound, images and, in general, any single or multiple, sequential, or concurrent, human-computer input, output, or action."[54]

An NSF program manager in this period confided that his proudest accomplishment was doubling the already ample funding for natural language understanding. Despite decades of research, speech and language recognition were not commercial successes. Use was rarely discretionary; at best, it assisted translators, intelligence analysts, and people trapped by a phone answering system or whose circumstances limited keyboard use.

8.3   INFORMATION SYSTEMS EXTENDS ITS RANGE

GUIs were not quickly adopted by organizations, but spreadsheets and business graphics (charts and tables) were important to managers and became foci of information systems research. Remus (1984) contrasted tabular and graphic presentations. Benbasat and Dexter (1985) added color as a factor, although color displays were rare in the 1980s. Many studies contrasted online and paper presentation because most managers worked with printed reports. Research into individual cognitive styles was abandoned; the concept of cognitive fit between task and tool was introduced to explain apparently contradictory results in the technology adoption literature.[55]

---

[53] Roland and Shiman, 2002.
[54] National Science Foundation, 1993.
[55] Cognitive styles research ended after a pointed critique (Huber, 1983); for cognitive fit, see Vessey and Galletta, 1991.

In 1986, Jane Carey initiated a series of symposia and books titled *Human Factors in Information Systems*.[56] Topics included design, development, and user interaction with information. As corporate adoption of minicomputers and intranets matured, studies of email and other collaboration support appeared. The focus shifted away from maximizing computer use—screen savers had become the main consumer of processor cycles.

Hands-on managerial use remained atypical, but group decision support system (GDSS) research sought to change that. It emerged from work on decision support systems that were designed for an executive or a manager. Central to GDSS was technology for meetings, including brainstorming, idea organization, and online voting features. The standalone systems were initially too expensive for widespread group support; hence, the focus on "decision makers" and research in schools of management, not computer science departments or software companies.[57]

As computing costs dropped and client-server local area networks spread in the mid-1980s, more laboratories built meeting facilities to conduct research.[58] GDSSs morphed into "group support systems" that included support for non-managerial workers, catching the attention of CHI researchers and becoming a major IS contribution to Computer Supported Cooperative Work (CSCW), discussed in the next section. In 1990, three GDSSs were marketed, including a University of Arizona spin-off and an IBM version that licensed the same technology. The systems did well in laboratory studies but were generally not liked by executives and managers, who felt that their control of meetings was undermined by the need for a professional meeting facilitator, among other things.[59] Executives and managers were discretionary users whose involvement was critical; the products were unsuccessful.

With the notable exception of European sociotechnical and participatory design movements, non-managerial end user involvement in design and development was rare, as documented in Friedman's (1989) comprehensive survey and analysis. Then Davis (1989), having been exposed to CHI usability research, introduced the influential Technology Acceptance Model (TAM). A managerial view of individual behavior, TAM and its offspring focused on perceived usefulness and perceived usability to improve "white collar performance" that is "often obstructed by users' unwillingness to accept and use available systems." "An element of uncertainty exists in the minds of decision makers with respect to the successful adoption," wrote Bagozzi et al. (1992).

TAM is the most cited HCI work that I have found in the IS literature. Its emphasis on the *perception* of usability and utility is a key distinction. CHI *assumed* utility—consumers choose technologies that they believe will be useful and abandon any that are not—and *demonstrated* usability improvements. TAM researchers could not assume that an acquired system was useful. They

---

[56]   E.g., Carey, 1988.
[57]   Kraemer and King (1988) and Nunamaker et al. (1997) summarize decades of GDSS research.
[58]   Begeman et al., 1986; DeSanctis and Gallupe, 1987; Dennis et al., 1988.
[59]   Dennis and Reinicke, 2004.

observed that *perceptions* of usability influenced acceptance and the likelihood that it would become useful. CHI addressed usability a decade before TAM, albeit actual usability rather than perceived usability. In CHI, perception was a secondary "user satisfaction" measure. The belief, not entirely correct, was that measurable reductions in time, errors, questions, and training would eventually translate into positive perceptions. "Acceptance," the "A" in TAM, is not in the CHI vocabulary: discretionary users adopt, they do not accept.

The IS and CHI communities rarely mixed. When CHI was more than a decade old, the September 1994 issue of *Harvard Business Review*,[60] a touchstone for IS researchers, published "Usability: The New Dimension of Product Design." The article did not mention CHI at all and concluded, "user-centered design is still in its infancy."

8.4   COLLABORATION SUPPORT: OIS GIVES WAY TO CSCW

By the late 1980s, three research communities were addressing small-group communication and information-sharing:

1. OA/OIS;

2. IS group decision support; and

3. CHI researchers and developers who looked beyond individual productivity tools for "killer apps" to support emerging work groups connected by local area networks.

OA/OIS had led the way, but it was fading as the minicomputer platform succumbed to competition from PCs and workstations. The concept of "office" was proving to be problematic; even "group" was elusive: organizations and individuals are persistent entities with long-term goals and needs, but small groups frequently have ambiguous membership and shift in character when even a single member comes or goes. In addition, people in an organization who need to communicate are often in different groups and fall under different budgets, which complicated technology acquisition decisions at a time when applications were rarely made available to everyone in an organization.

The shift was reflected in terminology use. First, "automation" fell out of favor. In 1986, ACM SIGOA shifted to SIGOIS and the annual AFIPS OA conference was discontinued. In 1991, "office" followed: *Transactions on Office Information Systems* became *Transactions on Information Systems*; *Office: Information and People* became *Information Technology and People*; and ACM's Conference on Office Information Systems became Conference on Organizational Communication Systems (COOCS, which in 1997 became GROUP).

---

[60]   March, 1994.

The AI summer of the 1980s, a key force in the rise of OA/OIS, ended with the failure of DARPA's multi-billion dollar Strategic Computing Initiative. Few offices were automated by 1995. CHI conference sessions on language processing had diminished earlier, but sessions on modeling, adaptive interfaces, advising systems, and other uses of intelligence increased through the 1980s before declining in the 1990s. As funding became scarce, AI employment opportunities dried up and conference participation dropped off.

In 1986, the banner "computer supported cooperative work" attracted diverse researchers into communication, information sharing, and coordination. Building on a successful 1984 workshop,[61] participants came primarily from IS, OIS, CHI, distributed AI, and anthropology. Four of the 13 program committee members and several of the papers were from schools of management.

CSCW coalesced in 1988 with the publication of Greif's *Computer-Supported Cooperative Work*. SIGCHI launched a biennial North American CSCW conference. With CSCW reluctant to meet in Europe, a European CSCW series (ECSCW) began in 1989. With heavy participation from telecommunication and computer companies, North American CSCW focused on small groups of PC, workstation, and minicomputer users. Most were within an organization; some were linked by ARPANET, BITNET, or other networks. European participation was primarily from academia and government agencies, focused instead on organizational use of technology, and differed methodologically from North American IS research. Initially, the division was significant. When *Computer Supported Cooperative Work: An International Journal* appeared in 1992, all of its editors were in Europe.

The pace of change created challenges. In 1985, successful use of technology to support a team was a major accomplishment; in the early 1990s, applications that provided awareness of the activity of distant collaborators were celebrated. By 1995, the internet and web had arrived and *too much* visibility of remote activity could raise privacy concerns and create information overload. Other phenomena were no sooner identified than they vanished. A discovery that IT investments were not returning benefits, called the "productivity paradox," drew attention in 1993, but five years later the same researcher reported that IT investments spurred productivity growth.[62] Later, a widely reported national study that showed adverse health effects of internet use on young people, called the "internet paradox," was replicated a few years later by the same team, and the effects were gone.[63]

European and North American CSCW gradually aligned as European organizations acquired the commercial software products that were built and studied in North America, and North Americans discovered that the ECSCW organizational context focus was often decisive when deploying tools to support teams. Organizational behaviorists and social theorists remained in

[61]   Described in Greif, 1985.
[62]   Brynjolfsson, 1993; Brynjolfsson and Hitt, 1998.
[63]   Kraut et al., 1998, 2002.

their home disciplines, but ethnographers, who by studying technology use in developed countries were marginalized in traditional anthropology departments, were welcomed by both CSCW and ECSCW. Perfectly positioned to help bridge these groups was the influential branch of Xerox PARC established in 1987, in Cambridge, UK. EuroPARC comprised strong technologists, sociologists and ethnomethodologists who collaborated with a range of European researchers and presented at CSCW and ECSCW conferences. Silicon Valley corporate entrepreneurship meets government-sponsored European research.

Despite the challenges of building on sand that was swept by successive waves of technology innovation, CSCW continued to attract a broad swath of HCI researchers. Content ranged from highly technical to thick ethnographies of workplace activity, from studies of instant messaging dyads to scientific collaboratories involving hundreds of people dispersed in space and time.[64]

Desktop videoconferencing was a signature CSCW and IS research field. Video phones were envisioned by Alexander Graham Bell and first realized in the 1930s, and expensive, unsuccessful products appeared in the 1960s in Sweden and the U.S. As telecommunication companies sought to promote bandwidth use, Bellcore (a successor to Bell Labs following the breakup of AT&T) and several regional telecoms undertook desktop video experiments. Xerox had launched an office automation product line, and Xerox PARC experimented with videoconferencing at sites in Palo Alto; Portland, Oregon; and Cambridge, UK. The language and keyboard neutrality of videoconferencing led to a strong international appeal. Influential research was undertaken in Japan at NTT and Keio University. Preparations in the early 1990s for moving Germany's capital from Bonn to Berlin inspired Desktop videoconferencing research at the prestigious GMD national research labs. The University of Toronto explored desktop conferencing interfaces. Early scientific collaboratories featured video links. CSCW conferences were the major venue for publishing this work, attracting researchers whose native language was not English.

In 1986, a *Management Science* article by Daft and Lengel proposed "Media Richness Theory," postulating that adding video to business communications would aid decision-making effectiveness in situations lacking clarity. This led to extensive experimental work in the IS field. Some studies supported the theory; some did not. These were not published in CSCW. Much as human factors researchers left CHI, most IS researchers left CSCW in the early 1990s. IS submissions to CSCW that bristled with acronyms and terminology unfamiliar to SIGCHI reviewers were rejected.

The IS approach to studying teams was shifting to organizational behavior from social psychology; the latter remained in favor in CSCW. The organizational focus conflicted with the computer and software company focus on the context-independent small-group support favored by many social psychologists. Splits based on paper rejections are not amicable. An IS newsletter, *Groupware Report*, omitted CSCW from its list of relevant conferences. The Hawaii International Conference on System Sciences (HICSS) became a major IS pre–journal publication venue for

[64]   Olson and Olson (2012) is a handbook chapter reviewing CSCW collaboration technologies.

group support research. Some IS researchers participated in COOCS and a Groupware conference series initiated in 1992, comprising a trade show with research presentations and a proceedings.

8.5   PARTICIPATORY DESIGN AND ETHNOGRAPHY

European efforts to involve future users in design, discussed in Chapter 5 (1965–1980), achieved wider attention in this period. Their focus remained on systems developed in organizations for mandatory use, not on "shrinkwrap" or commercial software. Sociotechnical design took a managerial perspective, with user involvement intended both to improve functioning and increase acceptance. The union-based Nordic participatory or cooperative design approach was more focused on empowering future hands-on users of the completed system. Both involved intense, carefully planned, and protracted collaboration between the design and development team and the future users.

It is not surprising that these approaches influenced human factors and ergonomics (Rasmussen, 1986); as noted in Chapter 2, Lillian Gilbreth, considered by some to be the founder of human factors, had stressed the importance of consulting workers in designing non-discretionary work processes. More surprising was an outcome of a 1985 participatory design conference in Aarhus, Denmark. In 1988, Lucy Suchman published a widely read review of a book based on the conference.[65] Although participatory design critiqued standard top-down IS approaches to developing systems for non-discretionary users, it resonated with CHI researchers whose focus was on commercial applications and discretionary use. Very different contexts: why the rapport? Both had the goal of empowering hands-on users. Also, both sets of researchers were primarily baby boomers to the political left of the World War II generation that still dominated HF&E and IS.

Ethnography was another source of insight into potential users and contexts of use that drew attention in the late 1980s, following presentations of studies of workplace activity at CSCW'86 by the Xerox PARC team led by Suchman, as well as their published work. As the CSCW'88 program chair, Suchman lured Nordic researchers across the ocean to mix with North Americans. For several years thereafter, Participatory Design and CSCW conferences were collocated.

---

[65]   The book is Bjerknes et al., 1987; the review is Suchman, 1988.

The Shifting Focus of Interface Development

Figure 8.2: The principal locus of hands-on users' attention to the computer interface changed over time.

The evolution of HCI from the engineering and computer science perspective is reflected in Figure 8.2, taken from my 1990 paper, "The Computer Reaches Out." Engineers and operators initially interacted at the hardware level, connecting cables, setting switches, loading tapes, and so on. Software programs then became the dominant interaction medium. With terminals, the interface moved to the display and keyboard; in the late 1980s, with satisfactory perceptual and motor solutions in hand for many tasks, cognitive aspects moved to the fore. The fifth focus, support for group activity, was spotlighted by CSCW and grew in salience.

The figure does not include organizations. IS research started earlier with an organizational focus. A mainframe was acquired to meet enterprise goals. Eventually, minicomputers and networked microcomputers supported divisions and groups within organizations. As CHI moved from 4 to 5 in Figure 8.2, IS moved from the unseen 6 (organizations) to 5. IS and CHI researchers converged at the first two CSCW conferences.

The anthropomorphic metaphor of computers reaching out was intended to spur reflection and the realization that, in fact, the people and organizations described in this book did the painstaking work to extend the reach of the technology. The computer is not an alien life-form, it comes to reflect what we have learned about ourselves and how we work.

8.6   LIBRARY AND INFORMATION SCIENCE: TRANSFORMATION UNDER WAY

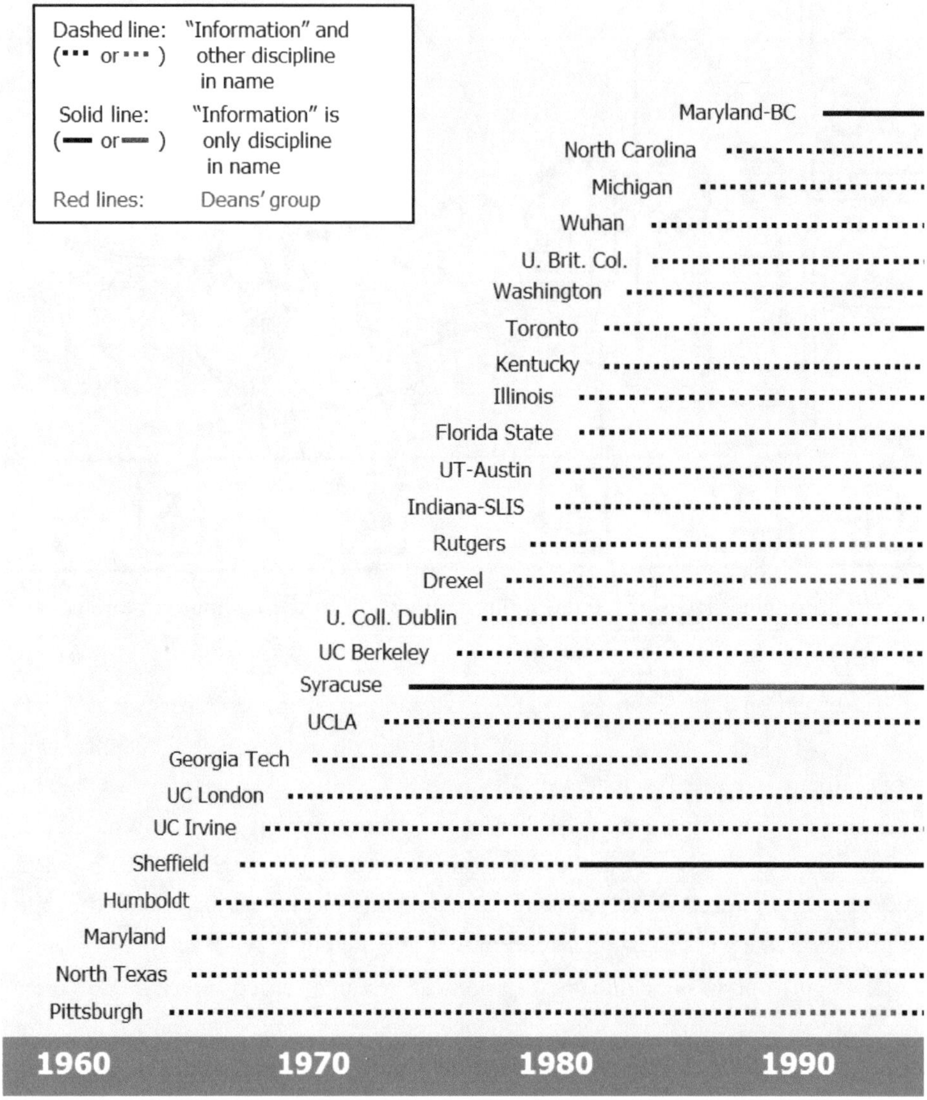

Dashed line: "Information" and
(••• or •••)   other discipline in name

Solid line: "Information" is
(— or —)   only discipline in name

Red lines: Deans' group

	1960	1970	1980	1990
Maryland-BC				
North Carolina				
Michigan				
Wuhan				
U. Brit. Col.				
Washington				
Toronto				
Kentucky				
Illinois				
Florida State				
UT-Austin				
Indiana-SLIS				
Rutgers				
Drexel				
U. Coll. Dublin				
UC Berkeley				
Syracuse				
UCLA				
Georgia Tech				
UC London				
UC Irvine				
Sheffield				
Humboldt				
Maryland				
North Texas				
Pittsburgh				

Figure 8.3: Early iSchools and when "Information" became part of their names.

The position of professional schools in research universities can be contentious, but those that bring in money or prestige are generally welcome. The prestige of library schools declined as libraries lost their monopoly on information. Between 1978 and 1995, 15 American library schools were shut down.[66] As seen in Figure 8.3, many survivors added "Information" to their names, usually by becoming Library and Information Science, through the 1970s and 1980s. By 1995 five were solely schools of Information. The humanities orientation was giving way as librarianship was changed by technology; IT staff salaries rivaled those of library professors.

Robert S. Taylor, a dean at Syracuse University, had led the way by changing the school name from Library Science to Information Studies in 1974. He proposed distinguishing information science and information engineering, and in 1980 initiated the first advanced degree in information resources management. In 1986, he published an influential, user-centered "value-added processes" model. His approach was more fully developed than CHI approaches of the time, but his "users" were primarily librarians and other information professional specialists. One of his six key criteria was cost saving, equated to "lower connect-time price," which reflected the mainframe orientation that remained dominant.

Change was not smooth. The closer a discipline is to pure information, the faster Moore's law and networks disrupt it once a tipping point is reached. Photography, music, news,…and libraries. The exclusion of information technology studies in library schools had once been understandable, given the cost and the limitations of early systems, but when useful systems arrived, there was little time to adjust. Young information scientists, their eyes fixed on a future in which past lessons might not apply, were reluctant to absorb a century of work on indexing, classifying, and accessing complex information repositories. Knowledge and practices that still applied had to be adapted or rediscovered. The conflicts were exposed in Machlup and Mansfield's landmark 1983 collection, *The Study of Information: Interdisciplinary Messages*. In it, W. Boyd Rayward outlines the humanities-oriented and the technological perspectives and argues that they had converged. His essay is followed by commentaries attacking him from both sides.

For several years starting in 1988, deans at Pittsburgh, Syracuse, Drexel, and Rutgers converged annually to share their approaches to explaining and managing multidisciplinary schools (the red lines in Figure 8.3). Despite this progressive effort, Cronin (1995) depicted LIS at loggerheads and in a "deep professional malaise." He suggested that librarianship be cut loose altogether and that the schools establish ties to the cognitive and computer sciences. The web had just appeared: more radical change lay ahead.

---

[66]   Cronin, 1995.

CHAPTER 9

1995–2005: The Internet Era Arrives and Survives a Bubble

In this decade, online activity became a global phenomenon with the full emergence of the internet and the web. A sharp increase in U.S. labor productivity has been attributed to both the production and the use of computers and communication technology. Mobile phones, especially those with messaging-friendly data plans, contributed to the spread of digitally-mediated communication. To sustain this growth, internet engineers did the equivalent of changing the engines of an accelerating plane as it flew, expanding the internet domain name space and increasing network capacity and reliability without interrupting service.

The decade also encompassed the rise and collapse of an enormous speculative technology-driven economic bubble. Its collapse was followed by a steady recovery that laid the groundwork for the solid advances that were to come.

9.1 THE INTERNET AND THE WEB

Most people experienced a seamless transition in 1995 when the NSFNET was decommissioned and the Commercial Internet Exchange (internet) came alive. NSFNET, managed by NSF, was created in 1985, a year after the Department of Defense severed its ties to the ARPANET. Wary of unruly ARPANET developers, it had set up a parallel MILNET, but continued supporting the ARPANET until 1990 to avoid disruption.

The 1995 transition was more than a name change. NSFNET "acceptable use policy" prohibited commercial activity, although ambiguity had arisen and enforcement had waned. With the web stirring, clarity was needed. The internet was now unequivocally open for business.

Prior to 1995, dial-in commercial online services existed: CompuServe, GEnie, Prodigy, and America Online (later AOL). The first two were initially efforts to obtain revenue from unused mainframe cycles. The cost per hour in the late 1980s was over $50 in today's dollars. By the early 1990s, rates remained high during business hours but dropped to around $10 per hour for evening use. Consumer use first centered on proprietary email and online forums. Other applications and subscriber-only content appeared over time. In 1996, AOL shifted from hourly to monthly fees and others followed. Services remained limited—a person could generally only interact with other subscribers to the same service and might be limited to 30 free email messages per month. Widespread predictions that proprietary, curated content would be more attractive than the "wild west" of the internet and world wide web fell flat. By the end of the 1990s, only AOL remained a significant presence. The success of the internet surprised many of those building it, who expected a commercial or government effort to replace it.[67]

For those lacking ARPANET access and unable to afford high rates, Usenet (and EUnet in Europe), Bitnet, and FidoNet transmitted files or messages from computer to computer to span great distances while minimizing long-distance charges. These platforms were primarily used for information exchange, not commerce. FidoNet connected the modem-based community bulletin boards (see Chapter 7, 1980–1985), which by the end of the decade had more participants than the online service providers. CSNET was a low-cost, dial-in system maintained by universities. By collectively negotiating low bulk rates from telecommunications companies, universities successfully resisted efforts to charge per byte when the internet era arrived. Usenet primarily relied on cycles donated by university computers. It supported topic-based newsgroups that were very popular, establishing, for the first time, large informal exchanges of geographically distributed people who never met. Traffic peaked each year when new students arrived and tried the system and then dropped off. The "endless September" of 1993 occurred when AOL connected to Usenet. Traffic peaked—and kept climbing. The general public contributed large volumes of pornography, pirated software, and other material that universities didn't want to host. Traffic eventually shifted to the internet, but experiences with Usenet and the other platforms influenced the way people approached the commercial internet.

Mosaic and Netscape web browsers were released in 1993 and 1994, and the HTTP communication protocol in 1996. They were critical to the success of the web, which became a frontier into which speculators poured like homesteaders in the Oklahoma Land Rush. Some homesteaders in Oklahoma knew how to farm, but only platform vendors such as Sun Microsystems knew how to make money from the web in the mid-90s, although that did not stop people from trying. An early obstacle was the need to work out secure web-based credit-card transactions, which had not arisen in the non-commercial era.

9.2   COMMUNICATION, COLLABORATION, AND COORDINATION

Prior to the internet and web, computer and telecommunication support for group activities had largely been explored within organizations. The U.S. government had subsidized computer innovation in universities, but the costs were exorbitant—estimates of the real cost of sending an email message ranged from a few dollars to hundreds of dollars in the 1960s. Now, distributed activities were possible over the internet or among organizations.

---

[67] Kowack, 2008.

The most-cited CSCW paper, Resnick et al. (1994), described an architecture for recommender systems and its use in recommending Usenet newsgroups. Recommenders quickly became an active field of research and application. Early systems suggested films to watch or music to listen to, drawing on collaborative filtering of prior choices of other people and content filtering to find associated items. Recommender systems suggest people you might contact on social media sites or products you might consider. Recommender system research is featured in an unusual range of conferences, as it draws on HCI, information retrieval, data mining, and machine learning,

NSF and National Institutes of Health (NIH) provided substantial support in this period for ambitious scientific and engineering collaboratories linking researchers around the world on topics such as atmospheric science, earthquake simulation, and medical diagnosis, reviewed retrospectively in Olson et al. (2008). Technical and behavioral factors identified in these efforts could have broader application when technology costs dropped. They found that asynchronous communication and sharing was more important than real-time video, a technology that was initially emphasized in some collaboratories. Face to face meetings proved to have high value when they could be arranged.

In addition, computer-mediated human-human interaction was amplified by the spread of mobile phones and computers. Text messaging took off from 1995 through 2000, initially with short message services (SMS) supported by some telecommunications companies. Where SMS was a significant added charge, internet-based computer messaging was an option for people already paying for internet access: ICQ, AOL Instant Messenger, Yahoo! Pager, and MSN Messenger were launched in successive years from 1996 to 1999. China's QQ service arrived in 1999. Mobile phone buddy lists that appeared in this period also laid the groundwork for social networking, which had an early entrant with Classmates.com in 1995 but became prominent a decade later.

The web inspired new communication tools. The term "wiki" was coined in 1995, followed two years later by "weblog," or blog. In 1999 two successful blog platforms were released, LiveJournal and Blogger. However, wikis and blogs were niche activities as the century came to an end; in contrast, messaging spread rapidly, especially among young people who had computer or mobile phone access.

9.3   THE BUBBLE AND ITS AFTERMATH

Computer enthusiasts, accustomed to being outside the mainstream, found it odd when internet addresses and website URLs began popping up in comic strips and on billboards. From 1995 to 2005, most people in developed countries, and many outside them, first used computers and got online. Internet use went from 15 million people to over a billion. The stage was set in 1993, when non-nerdy use of computers was crucial in resolving crises in four blockbuster movies: *The Fugitive*, *The Firm*, *Sleepless in Seattle*, and *Jurassic Park*—skilled use by doctor Harrison Ford and attorney Tom Cruise, and by small girls booking an airline ticket and stopping velociraptors.

Services used proven technology to exploit rising bandwidth, processor power, digital memory, and consumer markets and enter new fields such as digital photography, digital books, and digital music. Napster (launched in 1999) and Pandora (2000) provided music-sharing technology similar to Usenet and other file-sharing predecessors. Sharing that had been tolerated was now on a scale that threatened publishers. The U.S. Digital Millennium Copyright Act of 1998 implemented two 1996 treaties of the United Nations World Intellectual Property Organization. Debates and lawsuits spring up in the early 2000s, as they had for software code in the 1970s and "look and feel" interface issues in the late 1980s. Getty Images sued to protect copyrighted photographs, the Author's Guild sued Google over book digitizing, and Metallica and later the Recording Industry Association of America sued Napster.

Not all potential beneficiaries of improved hardware capabilities thrived. Active Worlds launched in 1995, but people found little to do in virtual worlds other than chat and play constrained games. Real-time multicast backbone (Mbone) videoconferencing gathered steam in the mid-1990s; an ambitious November 1994 multicast of a Rolling Stones concert was overwhelmed, but less ambitious events succeeded. If you are unfamiliar with Mbone, it is because Mbone development ceased when the web consumed most of the oxygen in R&D labs. The web was more reliable, required less effort, and had a mind-boggling potential audience. Eventually support for real-time communication resumed, as reflected in the launch of Skype in 2003 and O'Reilly Media's successful Web 2.0 conference in 2004.

The web also curtailed AI efforts to build powerful knowledge-based systems. With information on almost any topic accumulating and available on the web through search, the effort to embed knowledge in application software lost its appeal. Moving product support online is a simple example; more ambitious efforts to embed knowledge in systems were undertaken into the early 1990s. In contrast, the rising volume of online information and activity encouraged explorations of statistical and machine learning approaches. A new crop of ultra-intelligent forecasts bloomed.

Entrepreneurs and investors sensed accurately that there was gold out there, but they were wildly optimistic about where it would turn up and how quickly it could be mined. Some pioneers prospered: Amazon and Netflix established successful trading posts in 1995 and 1997, respectively, both initially taking orders online and sending physical content by postal mail. However, myriad new ventures with no realistic business plans attracted investors and went public. The valuation of major technology companies reached unimaginable heights. In January 2000, in a mind-boggling move, AOL bought Time Warner. Time Warner comprised not only Time Magazine and Warner Brothers studios, but also HBO, CNN, DC Comics, New Line Cinema, Turner Broadcasting System, Castle Rock Entertainment, Cartoon Network Studios, and more. All of this was now owned by an internet service provider.

Did people digest the sale and conclude that speculation was out of hand? Did the January 1, 2000 failure of the Y2K disaster to materialize—widely hyped predictions that computers not programmed to handle dates beyond 1999 would crash and create chaos—signal the fallibility of tech predictions? In any case, the stock bubble peaked two months later. As internet companies crashed out of business, the NASDAQ index dropped from over 5,000 to just above 1,000.

The tide went out, carrying away investments and the websites that people hadn't visited. But it left behind hundreds of millions of people who had acquired computers, internet access, and new skills. Looking for useful technologies, they found them when seeds that were planted during the bubble germinated.

In 2000, Google, a new search engine provider that had larger competitors, devised a new business model: advertisements keyed to search terms attracted users and provided revenue. By the time Google went public in 2004, developers and investors had a clearer sense of technology opportunities. The wiki concept preceded the crash; Wikipedia started in 2001, with 20,000 articles by the end of the year, and continued growing. Two platforms launched in 1999 were adopted by legions of bloggers in the early 2000s: In 2003, LiveJournal hosted its millionth blog and Google acquired Blogger. Wordpress was launched the same year, and *Communications of the ACM* published a special issue on "The Blogosphere" in 2004.

Other communication technologies thrived. Aspects of mobile phone buddy lists, SMS, Instant Messaging, and the web merged in social networking sites. Joining alumni-oriented Class-mates.com in 2002 was Reunion.com, the same year Plaxo and Friendster launched as popular social networking sites. MySpace and LinkedIn launched in 2003, Flickr and Orkut in 2004, YouTube and Bebo in 2005. In 2006, Twitter was launched and Facebook went from restricted to open membership.

All of these were very successful for a time. In 2006, MySpace reportedly surpassed Google as the world's most-visited website. Other sites that emerged in this short timespan established followings in specific regions: Cyworld (Korea), Hi5 (Latin America), XING (Germany), and Viadeo (France). Yahoo! 365 and MSN Spaces were less successful entries that nevertheless validated the significance of social networking.

In summary, despite the disastrous, rapid failures between 1995 and 2000, the years that followed saw the rise of iconic internet companies that were still thriving a decade later.

The bubble impacted the four HCI fields differently. Human factors was the least disrupted: the internet and web were not initially a strong concern of government and established business systems administrators, and the early web revived a form-driven interaction style familiar to human factors. Internet-savvy CHI researchers were diverted but not disrupted. CHI focused as before on the interfaces, interaction models, and research methods for products that were enjoying early mass-market adoption, many of which were now internet and web applications. The rapid expansion of interaction opportunities and challenges led to the hiring of HCI professionals in computer science departments and in industry.

Information systems was suddenly in the spotlight in schools of business and management. Highly valued technology companies with little-understood methods turned the financial world upside down. The attention to IS had come at the expense of traditionally dominant disciplines such as marketing and finance, which reasserted themselves when the unpredicted collapse of the bubble revealed a lack of IS insight into technology value. Information science was also disrupted at this dawn of an information age, as hundreds of millions of people began accessing a rapidly expanding world of online information. Let's first address the two disciplines that were most affected.

9.4   THE FORMATION OF AIS SIGHCI

IT professionals and IS researchers had focused on the internal use of systems in organizations. The internet created more porous organizational boundaries. Employees downloaded instant messaging clients, music players, web apps, and other software, despite management concern about potential negative effects on employee productivity, and the IT staff's concerns about security. Facebook, Twitter, and other applications and services could be accessed in a browser without a download. Experience at home increased employee impatience with poor software at work that was often developed by the employer for mandatory use without the competitive pressure of commercial products that were chosen for use at home. Managers who had been hands-off users became late adopters; as they retired, younger replacements had hands-on experience. More executives as well as managers became hands-on early adopters of tools.

Significant as these changes were, the web had the more dramatic effect. Suddenly, organizations had to create web interfaces to customers and external vendors. Discretionary users! Although the collapse of the bubble in 2000 revealed that IT professionals and IS experts had not understood web phenomena well, Amazon and Netflix continued to thrive, demonstrating that new PC owners sought online sales and service. Business-to-business opportunities also beckoned. As the web became an essential business tool, IS researchers faced discretionary use issues that CHI had confronted 20 years earlier, whether they realized it or not.

Some realized it. In 2001, the Association for Information Systems (AIS) established the Special Interest Group in Human–Computer Interaction (SIGHCI). (Note the distinction between this and ACM's 18-year-old SIGCHI.) The founders' outline of HCI cited 12 CHI research papers.[68] They declared that bridging to CHI and information science was a priority. The charter of SIGHCI included a broad range of issues, but early research emphasized interface design for e-commerce, online shopping, online behavior "especially in the internet era," and the effects of web-based interfaces on attitudes and perceptions.[69] SIGHCI sponsored special issues of journals;

---

[68]   Zhang et al., 2004.
[69]   Zhang, 2004.

eight of the first ten papers covered internet and web behavior. As discussed in Chapter 11, the bridge to CHI did not materialize, but SIGHCI thrived.

9.5   DIGITAL LIBRARIES AND THE RISE OF INFORMATION SCHOOLS

For libraries to do more than digitize card catalogs, the cost of digital memory had to drop. An early PC could not come close to holding a single digital photo taken today on a camera or phone. And memory that existed was expensive! In 1980, Shugart (now Seagate) released a 5 MB hard drive that cost US$4,500, adjusted for inflation. A 10 MB version was in the 1983 IBM XT, the first PC that came with an internal hard drive. By 1990, comparably priced drives held over 300 MB. Despite that capacity increase, people had insufficient memory to save all of their email; collections of digital photographs, books, or audio were inconceivable. Nevertheless, storage capacity was on track to increase exponentially; digital information was destined to become significant. A new hard drive type, Integrated Drive Electronics, led to a sharp drop in prices in 1995, with a standard drive holding over 1 GB. By 1995, courses on digital indexing and retrieval technology were prerequisites for librarianship. Innovative research was not keeping pace with the changes in professional training in library and information schools,[70] but that would soon change.

In the early 1990s, Gopher, a structured system for indexing, searching, and retrieving documents, attracted attention as a repository for digital materials. H. G. Wells' (1938) "world brain" seemed to be within reach. Then the web hit, accelerating the transformation of information distribution. This was bad for Gopher but good for the library and information science research community. It was galvanized by "digital library" research and development awards totaling close to US$200 million from 1994 to 1999, jointly sponsored by NSF, DARPA, NASA, National Library of Medicine, Library of Congress, National Endowment for the Humanities, and the FBI. This was an unparalleled infusion of funds for a field rooted in the humanities. The historical ambivalence over the role of technology was ending. One digital library award went to Stanford, where in 1997 it supported Ph.D. student Larry Page as he developed the algorithm for ranking web pages that he used in co-founding Google the next year.

But what was this field? Cronin (1995) wrote that the focus of the information field was information access, spanning intellectual, physical, social, economic, and spatial/temporal factors. Acquired from sensors and human input, information flows through networks and is aggregated, organized, and transformed. Information routing and management is evolving, both within enterprises and across ever-more-permeable organizational boundaries. As personal information moves from often haphazardly organized shoeboxes of photographs and boxes of old papers to local or

---

[70] Cronin, 1995.

cloud-based digital formats, future accessibility can be even more unreliable. If not consciously and carefully managed, it can disappear in the blink of a system change.

In 2000, the ASIS appended "and Technology" and became ASIS&T. In ten years, the number of universities with a school or college in which "information" was the sole discipline in its name had increased from three to ten. The next year, meetings of information school deans began, modeled on those of the late 1980s described in Chapter 7. By 2004, 12 universities were participating.

The emergence of a field called information seemed under way. At Penn State and Indiana University, new schools of information were created. Berkeley severed ties with the library field. However, the schools differed considerably in their emphases. Many retained ties to librarianship and the ALISE conference, as well as to ASIS&T and its information science heritage. It was not clear how this would develop.

9.6   HF&E EMBRACES COGNITIVE APPROACHES

David Meister, the author of *The History of Human Factors and Ergonomics* (1999), stressed the continuity of HF&E in the face of technology change:

> Outside of a few significant events, like the organization of HFS in 1957 or the publication of Proceedings of the annual meetings in 1972, there are no seminal occurrences . . . no sharp discontinuities that are memorable. A scientific discipline like HF has only an intellectual history; one would hope to find major paradigm changes in orientation toward our human performance phenomena, but there is none, largely because the emergence of HF did not involve major changes from pre-World War II applied psychology. In an intellectual history, one has to look for major changes in thinking, and I have not been able to discover any in HF.[71]

The origin of human factors in aviation psychology and its close ties to government systems helps explain this. The planning cycles for new military ships or jets project ahead 40 years. This leads to an emphasis on refinement over time, as well as complicating the planning of semiconductor elements.

As the internet and web were capturing much of the world's attention, NSF HCI program managers in this period emphasized other topics. Their focus on speech and language recognition continued unabated: even after a separate Human Language and Communication Program was established in 2003, speech and language remained a major area in the HCI and Accessibility programs, and drew additional support from the AI program. "Direct brain interfaces" or "brain-computer interaction" technologies received attention, although they were not destined to reach discretionary home and office contexts anytime soon. CHI conferences, which in this period had

[71] Meister, personal communication, September, 2004.

very little on speech, language, or direct brain interaction, were rarely attended by NSF program managers. An NSF oversight review committee noted that a random sample of NSF HCI grants included none by prominent CHI researchers.[72] Despite the strong support, little progress in these areas was achieved in this decade.

Meister overlooked one discontinuity. In 1996, a Cognitive Engineering and Decision Making technical group formed and soon became the largest in HFES. This marked an interesting evolution: as noted in Chapter 7, CHI formed a decade earlier in part because of strident human factors opposition to cognitive approaches. The term "cognitive engineering" was first used by an early CHI researcher.[73]

In an equally surprising reversal, a thriving Human Performance Modeling (HPM) technical group formed in HFES in 1995. Card et al. (1983) had introduced human performance modeling explicitly to reform the discipline of human factors from the outside. But the effort logically belonged within HF&E because it shared the focus on non-discretionary expert performance. HPM was led by Wayne Gray and Dick Pew, who had been active in CHI in the 1980s. The last major CHI undertaking in performance modeling was a special issue of *Human–Computer Interaction* in 1997, but HPM was active in HFES 20 years later.

9.7 CONSUMER USE MUSHROOMS AND CHI EMBRACES DESIGN

As applications mature and use becomes routine, fine-tuning often becomes a human factors task. Email and text editing or word processing were in use by academics and computer professionals by 1995 and by much of the developed world by 2005. Papers on these topics no longer appeared at CHI, which had moved on to new discretionary use frontiers. Over the decade, research focused progressively more on information visualization, web design, smaller devices (digital pen, mobile handheld, tangible UI), and computer use outside of offices (home, automobiles, robots, public displays). The novelty of computer use outside organizations was marked by paper session titles heralding computer use "in everyday life."

Early in the decade successful computer-supported collaboration was an achievement; as noted in the previous chapter, concerns were later raised about privacy, interruptions, and multitasking that arose from too much access and visibility. Although some still aspired to build cognitive theory to help design,[74] cognitive methods and modeling slowly gave way to the social and ethnographic methods that were better suited to exploring and assessing activities that involved two or more people. Although specific low-level pursuits changed, continuity was evident at a higher level as explorations continued into multimedia, input devices, communication channels, and design methods.

[72]   National Science Foundation, 2003.
[73]   Norman, 1982; 1986.
[74]   Barnard et al., 2000; Carroll, 2003.

To chart the tumultuous decade, I examined CHI conference sessions for six years, 1995–1996, 2000–2001, and 2004–2005. To summarize, there was a shift from cognition and programming sessions to social and mobile sessions, with some constants: every year included multiple usability sessions. In 1995 there were three cognition and three programming sessions, but the next five years saw only one of the former and two of the latter. Social appeared first in 2001, with four sessions in 2004–2005, which also saw the only five sessions in the years studied that were devoted to privacy, multi-tasking, and attention interrupts. Handheld/mobile/wireless sessions first appeared in 2000–2001 and then proliferated.

The decade saw computer science departments hire HCI faculty, often somewhat grudgingly. To elevate the prestige of their conference papers, the CHI conference from 2002 to 2004 dropped its acceptance rate to 15%–16%. This succeeded; some faculty earned tenure in leading departments without publishing in a journal. It also eliminated most practitioner involvement. Smaller conferences sprang up as outlets for excluded material. For example, Usability Professionals Association (UPA) created a peer-reviewed paper track at this time. World wide web conferences included HCI papers from the outset. Accessibility was considered by many in CHI to be overly applied, with exceptions such as well-understood color blindness, and it was taken up in universal usability conferences in 2000 and 2003. SIGCHI membership declined throughout the decade.

9.7.1   DESIGN

For the psychologists and computer scientists of the early CHI community, interface design was a matter of science and engineering. They focused on performance and assumed that people eventually choose efficient alternatives, despite evidence dating back to cave paintings that humans highly value aesthetics. They left "entertainment" to SIGGRAPH with its high-priced machines.

Designers had participated in early graphical interface research.[75] Aaron Marcus worked full time on computer graphics in the late 1960s and William Bowman's 1968 book *Graphic Communication* influenced the design of the Xerox Star icons designed by Norm Cox.[76] For some time, though, the cost of digital memory and processing precluded artistic license, nicely documented in Fred Moody's 1995 ethnography *I sing the body electronic*. Just as computer graphics had split between those interested in HCI and those focused on photorealism in the 1980s, designers split into interaction designers and those focused on visual design. Both had roles to play, but visual design would thrive a decade later, just as photorealism would also eventually affect widely available devices.

In 1995, building on workshops at previous conferences, SIGCHI initiated the Designing Interactive Systems (DIS) conference. DIS aspired to be broad, but it drew more system designers than visual designers. And then an event in Silicon Valley altered the course of HCI. Steve Jobs,

[75]   Evenson, 2005.
[76]   Bewley et al., 1983.

fired from Apple when the Mac was failing in 1985, returned to a once-again-floundering Apple in 1997. He fired the company's HCI professionals and teamed up with designer Jonathan Ive. The success of the beautiful iMac launched in 1998 and the iPod in 2001 set in motion changes that are still being sorted out

In 2003, SIGCHI tried again, joining with SIGGRAPH and the American Institute of Graphic Arts (AIGA) to initiate the Designing for User Experience (DUX) conference series in San Francisco. DUX succeeded in establishing the significance of design to a large HCI community, with written research papers embracing visual and commercial design. Coordination across the three sponsoring organizations proved difficult; the final DUX conference was held in 2007. The magazine *ACM Interactions*, launched by CHI in 1994, steadily increased its attention to visual design over the years, both in content and stronger production values.

As computing costs dropped, the grip of the engineering orientation continued to weaken. CHI researchers eventually came around. Some labeled the study of enjoyment "funology" lest someone think they were having too good a time.[77] Even the NSF got in on the act, initiating a (short-lived) Science of Design program in 2003.

9.7.2   MARKETING

The roots of persuasion may be as deep as aesthetics, but marketing is often even more dimly regarded than design.[78] Marketing often pushed tech company engineers to rely on what marketers heard from customers about end-user preferences, a conduit that HCI professionals considered to be unreliable. However, designing websites to market products introduced complications. The mission of CHI professionals was helping "end users." Website owners want above all else a good consumer experience; everyone was aligned on this. However, consumers may want to quickly conduct business on a site and get off, whereas site owners want to keep users on the site to view ads or make additional purchases—the online equivalent of a supermarket placing common purchases far apart, forcing shoppers to traverse aisles where other products beckon. For individual productivity or entertainment tools, the customer and the user are one and the same. Here, they differed.

Further complications followed upon the reliance on metrics to measure success, especially for pages solely depending on advertising revenue. Marketing considered a boost in user's "time on page" or "click-through rate" to signal design success, whereas HCI professionals knew that it could indicate that users were confused or lost. Practitioners were left to work this out; CHI researchers largely avoided the dilemma: "Brandology" was not to follow design.

---

[77]   Blythe et al., 2003.
[78]   Marcus, 2004.

Group Norms Overcome Discretion

We exercise prerogative when we use digital technology, sometimes. Perhaps more often at home, less at work. The young and healthy have more choices than the aged and injured. We have no choice when confronted by that telephone answering system.

Software that was discretionary yesterday could be indispensable tomorrow. Collaboration forces shared conventions. Discretion evaporates when a technology becomes mission-critical, as word processing and email did in many contexts in the 1990s. When people exchanged printed documents, everyone could use a word processor of their choosing and decide whether to emphasize with italics or boldface. When they began co-authoring online, they generally had to adopt the same software and styles. Choice is only exercised collectively.

Shackel (1997) titled his progression "From Systems Design to Interface Usability and Back Again." Early designers focused on providing basic functionality; computer operators had to cope. The first work stations and especially PCs gave employees a choice—for example, which word processor to use, or even whether to stick with typing—and the design focus shifted to the discretionary human interface. Eventually, though, individual users became components in fully networked organizational systems, and discretion over some technology use was lost.

But it works both ways. Discretion increases when employees download free software, bring smartphones to work, and demand capabilities that they enjoy at home. When managers also use technology, they are less likely to mandate the use of one that is burdensome. For example, sophisticated language-understanding systems of this period appealed to military officers—until they themselves became hands-on users:

> I have seen generals come out of using, trying to use one of the speech-enabled systems looking really whipped. One really sad puppy, he said "OK, what's your system like, do I have to use speech?" He looked at me plaintively. And when I said "No," his face lit up, and he got so happy.[79]

In domains where specialized applications are essential and security concerns curtail openness, discretion can recede. However, a steady flow of new technologies that bring unanticipated, discretionary uses is ensured by semiconductor advances, competition, and the ease of sharing bits.

[79] Forbus, 2003; see also Forbus et al., 2003.

CHAPTER 10

2005-2015: Scaling

A powerful recovery from the aftermath of the bubble marked the decade starting in 2005. Computer and phone access increased from 1 billion people to about 3 billion. More work and play shifted online as we came to rely on online information sources, digital documents and photo management, social software, multiplayer games, online shopping and banking, and business-to-business transactions. These changes will play out for years to come.

It may seem paradoxical, but this triumph of technology was easy for technologists to underestimate. The world was catching up to where the tech-savvy had been. Computer science and other technically oriented academic departments grew steadily, as technology use expanded more rapidly in other departments. Membership in the professional organizations we have followed remained flat or even declined as relevant research and development diffused into other fields. Uncertainty about direction was evident at NSF, which terminated the Science of Design program and began and ended a program titled CreativIT.

Nevertheless, most HCI fields were buffeted by change; CHI focused on product categories that were gaining widespread use. Information was strongly affected as storage costs plummeted, transmission bandwidth increased, and cloud services blossomed. Communication Studies broadened its involvement in HCI; a discipline rooted in humanities and social sciences, it initially emphasized journalism, added radio, television, and mass media, and now brought a narrative perspective to computer-mediated communication, digital social media, and interaction design generally. Information systems were also affected as organization boundaries became more permeable to customers, vendors, and partners.

Health informatics became a significant focus in all HCI fields. Social media use in political crises and disaster response also drew strong media and researcher attention. The term crowdsourcing—for web-based recruitment of labor, ideas, or funds—was coined in 2005, the year Mechanical Turk was launched to farm out low-paid, short-term tasks to volunteers. The crowdfunding site Kickstarter was established in 2009.

Wikipedia, which like Linux was crowdsourced before the term existed, became controversial in 2005 when a study published in *Nature* put its reliability on a par with encyclopedias.[80] The authority of the publishing establishment had not been so challenged since Gutenberg. YouTube in 2005, Khan Academy and TED Talks that were available online in 2006, and Massive Open Online Courses (MOOCs) in 2008 cumulatively provided a phenomenal range of free, informative as well

80   Giles, 2005.

as entertaining online video. The educational playing field was transformed; but most secondary and higher education held fast to game plans designed in the chalkboard era.

10.1   CHI: THE ROAD FORKS

With billions of users, two paths under the CHI umbrella became pronounced.

1. The natural progression of CHI's longstanding ties to companies focused on mass-market applications led to major web companies such as Facebook and Google, to technology efforts directed at mass consumer markets such as games, and to issues such as privacy protection.

2. Organizations of all sizes in fields such as finance, real estate, store management, health care, online marketing, event management, and entertainment sought basic and applied research contributing to interaction design, adoption, and usability.

The largely academic CHI community of a few thousand people continued primarily along the first path, focusing on familiar products developed by large companies that hired their students as interns and employees. Few academics had expertise in other industries and often considered work in narrow domains to be inherently applied. The focus on understanding user behavior in specific domains shifted from CHI to organizations such as UPA. Of about 150 CHI 2015 sessions, "usability" appeared only in a tutorial course title and "UX" (user experience) in two session titles.

Different design and evaluation methods are suited for each branch. Heavily used web-based consumer applications can iterate rapidly with A/B testing—exposing multiple options to large numbers of randomly selected individuals. This is invaluable for companies such as Google, Facebook, Amazon, and Microsoft. A/B testing is less effective in many path 2 contexts. Design A could win an A/B test if most individual contributors in an organization favor it, but if managers or executives don't, it may not fly. Similarly, the appeal of collecting data from vast numbers of users is irresistible, although knowing which data to collect and how to analyze them is challenging.[81] Social media sites, search engines, and large digital repositories attracted external researchers interested in how they are used, how usage data can be analyzed, and apps that might be built to leverage them. Smaller endeavors were interested in revenue-generating advertising and marketing that are also integral to these sites, but that was of little interest to CHI and more a focus of IS research, covered later in this chapter.

The other major disruption in the mass market was Apple's phenomenal success, riding interest in the popularity of industrial design to become the world's most successful company. The iMac and iPod were followed by the iPhone in 2007 and iPad in 2010. Smartphones and tablets were

---

[81] Metrics generally reveal *what* happened and *when*, but not *why* or *how*, and *where* or *who* are potentially sensitive.

not new, but stylish design led to immense success and came to dominate aspects of usability and utility in many consumer technology companies. In the 1980s, visual designers were subordinate to interaction designers. In the 1990s, they were peers. By 2015, other HCI professionals often reported to designers in technology companies.

On the second path, domain-dependent applications benefit from wireframe prototyping, usability testing, and ethnographic approaches. Once a mainstay of computer science HCI classes, these skills are increasingly taught in design and communications programs. The latter teach web design, scripting, and video production without requiring math and programming classes that are not essential for many jobs. With the proliferating availability of design templates, professionals need not even master HTML5.

Job roles and titles were in turmoil. Roles were often shaped to fit a specific team or organization's context and need. The term "usability engineer" had for two decades provided an aura of solid respectability, but practitioners had progressively expanded their sets of tools and techniques. Many people in the tech industry saw usability narrowly as lab testing, and Apple's success without usability engineers was noticed. A new job title was needed to encompass the breadth of the work of professionals. Many were tried. In one large tech company the progression was Usability Specialist > Usability Engineer > User Researcher > UX Researcher > Design Researcher.

Overall, UX won. It was not a new term. In 1993, Don Norman took a position at Apple with the title User Experience Architect. The Design of User Experience (DUX) conference series began in 2003. In 2001, the Usability Professionals Association (UPA) had launched *User Experience* magazine, which became UX magazine in 2005. Articles in the magazine continued to favor "usability" through 2008. Belatedly, in 2012, UPA became UXPA, the User Experience Professionals Association.

However, there is no consensus definition. UX spans usability, qualitative approaches, data mining, interaction design, and graphic or visual design. DUX conferences were co-sponsored by CHI and the AIGA. UX professional appeared on lists of occupations in high demand, but even though companies knew that *something* was needed to fill the gap between writing code and delivering awesome features, they did not necessarily know *what* skills were key.

Large tech companies could afford to construct diverse teams of specialists, but smaller organizations looked for an individual who could do it all. This led empirical researchers and designers to take on more than they were trained for. The meme "UX Unicorn" appeared: a "mythical user interface design architect with an advanced adaptive skill range. Highly skilled in graphic design, prototyping wireframes, CSS3, HTML5, jQuery, multivariate testing, use case scenarios, marketing and branding."[82]

Unable to find anyone with all these skills, UX recruiters had to choose. "Interaction designer," often a product of design programs, fit in the developer-to-user gap. Many interaction

[82]   Owczarek, 2011.

designers and visual designers lack strong empirical skills, but with Apple's success revealing the opportunity of capitalizing on visual design unleashed by the drop in the cost of memory and processing power, UX professionals with visual and industrial design in their quiver had an edge.

The transition to handheld devices had taken time. Novel input devices and displays had to be designed and mastered, the reach of networks extended, and greater energy efficiency achieved. Major breakthroughs were Nokia's shift from business to entertainment phones in 2006, the Apple iPhone in 2007, and Android phones in 2008. Attention continued to shift from cognitive tasks to social behavior as social media use proliferated; there were 50 million Twitter tweets a day in 2010 and 500 million six years later. The PC was no longer center stage. A cornucopia of new hardware, software, and behavioral possibilities appeared: location awareness, technology use while driving, privately owned devices in schools and workplaces, and greatly amplified distractibility. Twitter streams were mined for any conceivable correlation, from post-partum depression to stock market forecasts. Graduate students swarmed over the Wikipedia mountain, with its complete public revision history. Mechanical Turk became both a tool and an object of study for researchers. A senior colleague returning from a conference mused, "I don't know what CHI is anymore."

Moore's law didn't wait. Semiconductors were embedded in greeting cards and magazine advertisements, and researchers embedded sensors and effectors in everything from clothes to contact lenses. But once again, power and networking limitations constrained the possibilities. The Internet of Things lies ahead.

Although CHI concentrated on widely used consumer applications, it showcased some domain-specific work beyond the long-standing interest in software development tools and practices. Crisis informatics grew from analyses of social media traffic. Health and fitness efforts gave rise to brain-computer interaction research. Studies linked stress to privacy, information overload, and multi-tasking. Researchers, including some not as young as they were in 1985, worked on interfaces for the aged. Accessibility studies blossomed, no longer restricted to applying vision science to help those who are color-blind. Sustainability support became another focus. The sentiment that CHI should avoid the humanities and focus on engineering and value neutrality diminished as the field embraced aesthetics and sought a larger role. Feminist theory and gender studies, critical theory, and humanistic perspectives extended CHI's academic reach.

The ease of forming web businesses affected CHI. The first wave of experimental psychologists was content to improve technology by identifying requirements and assessing prototypes, but the shift to computer science brought the dictum popularized by Alan Kay, "The best way to predict the future is to invent it." Contribution to theory did not match the appeal of contributing to a commercial startup in an era in which software marketing and distribution was easier, albeit still a challenge.

10.1.1 CSCW EXPANDS ITS CONSUMER FOCUS

After a decade of convergence, CSCW and ECSCW diverged again. In 1990, a wave of new consumer applications in North America had captured researcher attention there but not in Europe. In 2010, social media, multiplayer games, Wikipedia, and other web phenomena attracted CSCW students and academics while ECSCW remained focused on workplaces. As before, the rift could heal as the new applications established footholds in enterprises and their developers looked to better understand this market. Should the historically strong European centralized government support for research diminish, alignment could increase.

The name "Computer Supported Cooperative Work" was examined. Smartphones were not called computers; software went beyond support to primary roles in automating some tasks; more so than in the early days, it operated in situations of conflict and even coercion; and increasingly use was outside the workplace, including for games and entertainment. CSCW considered radical changes before extending the conference name to be "Computer-Supported Cooperative Work and Social Computing" in 2014.

Advances in processing power and graphics inspired new virtual reality efforts in ten-year intervals. AlphaWorlds emerged in 1995. Interest in Second Life virtual worlds earned it a *Business Week* cover story in May 2006. IBM and NSF created private Second Life islands on which to hold business meetings and review panels. Consumer knockoffs for children enjoyed a burst of popularity, such as Webkinz World in 2005. Second Life use peaked in 2009. IBM and NSF abandoned their islands. The next iteration was headset-based: HTC Vive, Microsoft Hololens, and Oculus Rift were announced in 2015 and released in 2016. Entertainment of all kinds is, of course, a major industry. Whether virtual and augmented reality find uses beyond entertainment and training remains an open question.

Don Norman and the Evolution of CHI

This is a fitting place to observe that the 35-year evolution of CHI through 2015 is reflected in the contributions of Donald Norman. A cognitive scientist at the University of California, San Diego, he founded a large academic HCI research group in 1980. He introduced the term cognitive engineering, and in the first CHI conference paper in 1983 applied engineering concepts, defining "user satisfaction functions" based on speed of use, ease of learning, required knowledge, and errors. He organized the 1985 collection, *User Centered System Design*, which had a long shelf-life. His influential *Psychology of Everyday Things* (1988) marked the shift to pragmatic usability from formal methods. Its 1990 reissue as *Design of Everyday Things* reflected a broader refocusing on invention. He worked at Apple as User Experience Architect and Vice President of Advanced Technology. In 2004, he stressed the role of aesthetics in *Emotional Design: Why We Love (or Hate) Everyday*

Things. In 2014, he became Director of the Design Laboratory at UC San Diego, working to steer design to approaches that place people at the center.

10.2   iSCHOOLS BECOME A WORLDWIDE PRESENCE

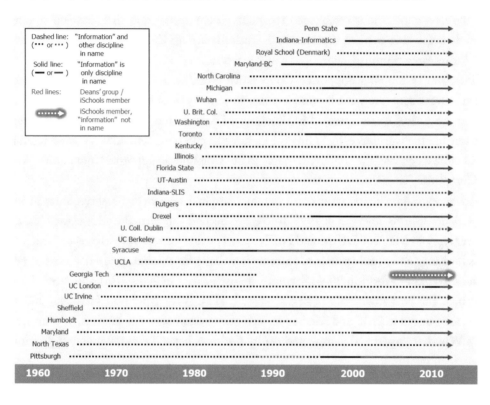

Figure 10.1: Ten years later: the shift to "Information" quickened and the iSchools banded together.

In 2005, the iCaucus of 12 dues-paying universities initiated an annual series of open iConferences, the first held at Penn State. The research conferences included a private meeting of the deans to discuss business and future conferences, share experiences, strategize on positioning iSchools, and consider possible joint initiatives. In 2005 and 2006, the halls and panels echoed with active discussions and disagreement about how the iSchools should position themselves. A key issue was whether the goal was formation of a new discipline or for each school to be a forum in which a set of pre-existing disciplines would collaborate, with the composition possibly depending on the history and context of each department.

The first possibility had been suggested earlier by Wersig (1992) and elaborated by Cronin (1995): Concepts around information might function "like magnets or attractors, sucking the focus-oriented materials out of the disciplines and restructuring them within the information scientific framework." The exponential growth in the production of digital information and the desperate need for theoretical and applied approaches to managing it suggested that if ever a brand new field might emerge suddenly, this could be it. Collectively, the iSchools had attracted, from top universities, luminaries in each discipline covered in this book.

However, many iSchool faculty considered themselves "a researcher in X working in an information school," where X could be library science, HCI, CSCW, IS, communication, education, computer science, or the like. The iCaucus had no journal and the iConference competed with conferences in these fields. Graduate students considering whether or not to identify with "information" noticed that iSchools primarily recruited faculty who made a mark in an established discipline, although over time iSchool Ph.D.s began landing tenure-track jobs at iSchools.

Figure 10.1 shows that by 2010, the number of "information" schools had tripled since 2005. (Not shown are Carnegie Melon and Singapore Management Universities, which joined the iCaucus despite having no academic units with "Information" in its name.) When the "iCaucus" had 21 member schools in 2009, Wiggins and Sawyer (2012) carefully analyzed their composition. Disciplinary representation varied, based more on local context than an intellectually derived plan. This paralleled how early cognitive science programs differed from university to university, one with a philosophy component, another with linguistics but no philosophy, some dominated by cognitive psychology, others more by computer science or AI. Several iSchools were dominated by a specific field other than library and information science; mainly computer science but also management, education, and communication. Although a few seemed content to be in the club and participated minimally in the iConferences, the dues and expectation that the dean would attend the annual meeting kept most engaged.

Overseas interest skyrocketed. Students crossed oceans to attend North American iSchools and new institutional members from overseas comprised a majority of the 70 admitted by 2015. Over 90% had "information" in their (translated) name. Professional organization loyalties remained splintered. Some retained strong ties to ALISE and library studies. ACM's SIGIR (Information Retrieval) conference retained its computer science orientation and high selectivity. The American Society for Information Science and Technology was a principal competitor. Based on individual membership, ASIS&T included some practitioners and overseas members. Reflecting the geographic trend, in 2013, the "American Society" became "Association," and its first overseas conference was scheduled.

Still unresolved was the key issue of whether a typical iSchool would be multidisciplinary (several groups working independently), interdisciplinary (groups with different primary allegiances working together), or a new discipline altogether. When cultures collide, traders form a pidgin language for basic communication; children who grow up speaking it creolize it into a new complete language. By analogy, would faculty from different fields work out a pidgin language, and assistant professors and graduate students whose first jobs are in Information Schools then creolize it into a language shared by all information professionals?

The uphill battle to achieve interdisciplinarity and a common language received a boost in 2013 with the publication of Robert Glushko's *The Discipline of Information*.[83] It identifies commonalities across disciplines, provides a wealth of compelling examples, and proposes a unifying terminology. In three years, it had course adoptions in over 50 universities.

10.3   INFORMATION SYSTEMS, UNDER PRESSURE, TURNS TO MARKETING

Information systems thrived during the Y2K crisis and the internet bubble; in response, other management disciplines—finance, accounting, marketing, strategy, entrepreneurship—acquired technical savvy. When the bubble burst and enrollments declined, the less well-defined IS niche faced a new challenge. The principal non-academic audience for IS researchers had been CIOs and IT directors, who focused on internal business functions. However, as one professor explained, "Most of the IT innovation, once only found in large scale organizations, has left the company." By relying on cloud services, vendors, and consumer applications, internal IT functions were outsourced. IT in organizations was increasingly administrative and outward-facing, managing vendors, customer-facing websites, and recruiting. Some IS groups in management schools were merged into other departments. Others reached out to corporate marketing directors, bringing IS into conflict with powerful management school marketing departments, many of which consider technology to be merely a tool for experts in consumer psychology.

SIGHCI remained the largest interest group in the AIS In 2009 it launched *AIS Transactions on Human–Computer Interaction*. In the third issue, Zhang et al.'s (2009) analysis of the IS literature from 1990 to 2008 documented the shift in the field from an organizational focus to one centered on the web and broader end-user computing.

At its inception in 2001, SIGHCI emphasized its commonality with ACM SIGCHI. Senior CHI and SIGHCI members attended each other's events in the early 2000s. Yet, the 2009 list of "related fields" did not mention CHI. The bridging effort had foundered. The dynamics that undermined efforts to bridge between CHI and human factors, OIS, and IS are explored in the next chapter.

Although the fields remained apart, IS researchers turned to the same datasets and technologies to measure customer behavior and attitudes, such as social media streams and fine-grained measurements of user activities. Following another path to gain respect or prosperity, some IS

---

[83]   Glushko, R., 2016, **now in its 4th edition.** http://disciplineoforganizing.org/.

leaders filed patents and considered start-ups, sometimes omitting details from papers to protect intellectual property.

10.4   TECHNOLOGY SURFACES IN HFES GROUPS

The historian David Meister's view that human factors is unwavering, quoted in Chapter 9 (1995–2005), was supported over the subsequent decade. The focus on slowly evolving government systems and processes is a factor. We may change our phone every year and our computer every 3, but aircraft carriers or power plants may take 10 years to build and last 40. However, within HFES, computer technology was once largely confined to one or two technical groups, and now it is relevant to most, as systems of all kinds look to human factors for efficiency gains. As other groups took on HCI activity, participation in the Computer Science Technical Group declined. The internet, Virtual Environments, Communications, System Development, and Augmented Cognition technical groups obviously focus on digital technology, and Health Care, Aging, Product Design, and Aerospace have significant HCI facets. Technical groups devoted to methods, such as human performance modeling and cognitive engineering and decision making, have broader, more theoretical charters, but contexts of application include technical systems.

In 2008, the 28-year-old NAS Committee on Human Factors was renamed the Committee on Human Systems Integration (COHSI). In 2010 it became a Board, reflecting an increase in requests from branches of government for human factors involvement and the conclusion that HCI was integral to the NAS. Despite the shedding of the name "human factors," the tie was unchanged: The first two Board of Human-Systems Integration (BOHSI) chairs went on to become presidents of HFES.

Some clouds appeared. The elevated prominence of BOHSI led to a challenge similar to that afflicting IS: other groups acquired in-house expertise, leaving less work for BOHSI. In 2013, HFES added "international" to the name of its annual meeting, concerned about losing international participation to local conferences, although travel restrictions for many members has kept the meetings in the U.S.

CHAPTER 11

Reflection: Cultures and Bridges

The major threads of HCI research interacted with each other sporadically but did not merge. It was not for lack of bridge-building efforts. The HFS co-organized the first CHI conference. CSCW sought to link CHI and IS. Mergers of OIS with CHI and CSCW were considered. AIS SIGHCI tried to engage with CHI. The iConference edged toward and then away from other fields, publishing its proceedings in the ACM digital library in 2011 and 2012.

Even within computer science, bridging is difficult. Researchers interested in interaction left SIGGRAPH for the CHI community rather than establish a bridge. Twenty years later, when standard platforms powerful enough for photorealism loomed, the DUX conferences sought to bridge these two fields and design but lasted only four years. SIGAI (previously SIGART) and SIGCHI cosponsor the Intelligent User Interface series, but participation in it has been outside mainstream HCI.

Many voices encourage interdisciplinarity and no one speaks against it, yet it is rare. We are now in a position to understand why it is difficult to achieve. In part, the overlap of focus was less extensive than it seemed. The major hurdles, however, were cultural, ranging from disciplinary priorities to linguistic and regional differences.

11.1 DISCRETION LED TO DIFFERENT METHODS

Human factors and management information systems turned to computers when discretionary hands-on use was uncommon. Similarly, LIS considered technology in the context of specialists for whom use was job-related. They all focused on training and rigorous experiments to squeeze out efficiencies in custom-made organizational systems. CHI occupied a new niche: discretionary use by non-experts. It focused on first encounters and casual use of mass-market software applications.

Some bridging occurred: HF&E and IS researchers shared journals for a time. For example, Benbasat and Dexter (1985) was published in *Management Science* and cited five *Human Factors* articles. However, bridging efforts to CHI failed. Researchers overestimated their shared interests. Had they been more conscious of this, areas of overlap might have been identified and focused upon.

11.2 DIFFERENT APPROACHES TO SCIENCE AND ENGINEERING

The scientific goal of generalization created tensions, especially for behavioral scientists. HF&E, with its primary focus on organizational systems in controlled settings, stayed close to the formal experimentation of science and engineering. IS, aware that no two organizations are the same within a single industry, much less across them, sought general, overarching theory. A general theory can be tested by any group. Many IS studies relied on teams of students recruited to be work groups, numerous enough to obtain statistical traction. Few real work groups are uniformly young with no history of collaborating or future prospects of doing so, but if the general theory is supported one can try to make a case. CHI researchers were often users of the consumer products on which they often focused, so they trusted their intuition when framing studies and were comfortable in shifting course as data came in. In contrast, many in the information field saw technology as a tool, not a focus of study. These differences obstructed appreciation and cross-publication. Human factors could be seen as focusing on the trees instead of the forest, IS as obsessing with hypothesis testing at the expense of ecological validity, and CHI as being atheoretical.

All three conducted experiments, influenced by psychologists trained in experimental hypothesis testing with human subjects. The discretionary/non-discretionary contrast is significant. Experimental subjects typically agree to follow instructions for an extrinsic reward: a reasonable model for nondiscretionary use, but not for discretionary use. CHI researchers relabeled "subjects" as "participants," which sounds volitional. Subjective reactions of discretionary users matter. CHI researchers therefore often ask for them. Unfortunately, lab study participants often tell experimenters what they want to hear, a phenomenon known as "demand characteristics." Newell et al. (1991) demonstrated it when testing a natural language understanding system. Half of the participants were told that the system had been real and half that it was a "Wizard of Oz" study with a human imitating software to be built later. When asked if they would like to have the system to use, most who believed it would be built later said "Yes" and those who believed it was available then and there said "No." Many published papers include glowing participant assessments of systems that did not survive.

The large design space drove a wedge between CHI researchers and practitioners. Researchers could compare two points in the space and obtain a statistically reliable difference with a moderate participant sample. Practitioners often found it effective to explore more points with smaller samples of each. Practitioner papers without a "p < .05" marker were rejected, even though sequences of pairwise comparisons are relied on by optometrists and others. Which is clearer, A or B? B or C?

The same goals apply everywhere—fewer errors, faster performance, quicker learning, greater memorability, and enjoyment—but the emphasis differs. For power plant operation, error reduction is critical, performance enhancement is good, and other goals are less important. For telephone order-entry, performance is critical and testing an interface that could shave a few seconds from a repetitive operation requires a formal experiment. In contrast, consumers often respond to visceral appeal and initial experience; avoiding obvious problems can be more important than striving for optimality. "Discount usability" and "cognitive walkthrough," less-rigorous methods with fewer participants, could be enough. Relatively time-consuming qualitative approaches, such as contextual design or persona use, can provide a deeper understanding when the context is critical and unfamiliar to the development team.[84]

Researchers from HF&E and theory-oriented IS were not impressed as CHI moved from its roots in scientific theory and engineering. The widely used "thinking aloud" method was based on the psychological verbal report technique of Newell and Simon (1972) and foreshadowed by Gestalt psychology. It was applied to design at IBM (Lewis and Mack, 1982) and broadly disseminated in Clayton Lewis's (1983) tutorial at the first CHI conference. Although unquestionably effective, it led some in other fields to characterize CHI as wanting to talk about experiences instead of doing research.

11.3   DIFFERENT PUBLICATION CULTURES

Table 11.1: The primary role of venues across disciplines

	Books	Journals	Conferences
Humanities	Polished work	Work in progress	Community-building
Sciences		Polished	Community-building
Computer Science (esp. in the U.S.)			Polished

As shown in Table 11.1, books are prized in the humanities and journal articles are the quality channel in most sciences. The disciplines of HF&E, IS, and LIS follow the sciences. Computer science diverged from this in the U.S. in the late 1980s, when conference proceedings became the final destination of most research. This followed a decision by ACM to archive proceedings once it was practical to assemble and publish them as low-cost, book-quality volumes prior to a conference (Grudin, 2011b). For research to enter the permanent literature, progressing to journal publication was unnecessary and could be considered self-plagiarism. In Europe, with no comparable professional organization and perhaps more respect for tradition or the value of more careful reviewing, the transition was slower. In Asia, where travel to conferences is more often out of reach, journal publication has remained essential.

---

[84]   Prominent examples: discount usability: Nielsen, 1989; cognitive walkthrough: Lewis et al., 1990; contextual design: Beyer and Holtzblatt, 1998; personas: Pruitt and Adlin, 2006.

This distinction became a major barrier to co-mingling. To justify treating conference papers as archival, acceptance rates were lowered. Computer scientists used this to convince colleagues from other disciplines that their conference papers were top quality. CHI generally accepts 20%–25% of submissions and dropped to 15% when it was moving into computer science departments. Most HF&E and IS conferences accept at least twice that proportion. I estimate that less than 15% of the work in CHI-sponsored conferences reaches journal publication, whereas an IS conference track organizer estimated that 80% of research presented there progressed to a journal.[85]

Human factors and information systems researchers told me that they stopped submitting to low-acceptance CHI and CSCW conferences because work polished to the level that they felt was required would be almost ready for journal submission, which their colleagues would value more. CHI researchers who attended inclusive, community-building conferences in other fields found much unpolished work, not realizing that by investing time in carefully planning session attendance, they could obtain an experience equal to that at their selective conferences. Having largely abandoned journals, CHI researchers may not seek out polished work in human factors or information systems journals, and other researchers who focus on the journal *Transactions on Computer-Human Interaction* will miss most work valued in that community.

Practitioners, discouraged by the high rejection rates and the disinterest in domain-specific work, sought "second-tier" conferences that mix research and best practices. In several interviews, senior researchers confided that they get more ideas from such venues than from major research conferences that favor methodological rigor over innovation. Despite numerous analyses that indicate that conference reviewing is not a reliable quality filter, academic CHI researchers have insisted on excluding most submitted work in order to argue that the surviving conference papers have merit.

Information schools have faculty on all sides of this divide. They report confronting the assessment issues and generally finding an accommodation, as computer science departments did earlier in discussions with other units. Open (free) access to published papers is discussed but has had less traction in this field than in mathematics and physics, where reviewing is much lighter, and in biology and medicine, where for-profit publishers control prestige journals and price access much higher than does ACM or IEEE.

11.4   VARIATIONS IN LANGUAGE, STANDARDS, AGE, AND REGION

Even within a single language, differences in terminology use impede communication. Where CHI referred to "users," HF&E and IS used the term "operators." In IS, a "user" was often a manager whose only contact with the computer was reading printed reports. For CHI, "operator" was demeaning, and users were always hands-on users. (In software engineering, "user" often meant "tool

---

[85]   Nunamaker, 2004.

user"—which is to say, developers.) Such distinctions may not seem critical, but they led to serious confusion and misconceptions when reading or listening to work from another discipline.

In HF&E and IS, "task analysis" referred to an organizational decomposition of work; in CHI, "task analysis" was a cognitive decomposition, such as breaking a text editing "move" operation into select, cut, position, paste. In IS, "implementation" meant organizational deployment; in CHI, it was a synonym for development. The terms "system," "application," and "evaluation" also had different connotations or denotations. Examples of significant misunderstandings that arose from not recognizing this are in the appendix.

Different perspectives and priorities were reflected in attitudes toward standards. HF&E researchers contributed to standards development, convinced that standards are integral to efficiency and innovation. Many in the CHI community believed passionately that standards *inhibit* innovation. In part, the attitudes reflected the different demands of government contracting and commercial software development. Specifying adherence to standards is a useful tool for those preparing requests for proposals, whereas compliance with standards can make it more difficult to differentiate a product. In general, people favor standards at system levels other than the level at which they are innovating, because such standards simplify their inputs and outputs. Both positions were defensible, but at the time, people on each side saw the other camp as obtuse. The positions converged when internet and web standards were tackled and more researchers developed software and saw the advantages of standard inputs and outputs.

CHI researchers who grew up in the 1960s and 1970s brought a generational divide to HCI. On balance, they did not share the prior generation's orientation toward military, government, and large businesses. Their appearance, dress, and demeanor was less formal. From the start, they reacted negatively to the lack of gender-neutral language that even now occasionally appears in the HF&E and IS "man-machine interaction" literature. Not until 1991 did the engineering-oriented Canadian Man-Computer Communication Society become the Canadian Human Computer Communication Society and, in 1994, the *International Journal of Man-Machine Studies* became the *International Journal of Human-Computer Studies*. Such differences diminished enthusiasm for building bridges and exploring literatures.

Challenges presented by regional cultures merit a study of their own. The strong presence in North America of the non-profit professional organizations ACM and IEEE led to developments not experienced elsewhere, including low-priced literature. Whether due to an entrepreneurial culture, a large unified consumer market, less government-driven research, or other factors, the success of software and web applications in the U.S. shaped the direction of research there. In Europe, government-sponsored research was central and favored organizational contexts. In the 1980s, some governments propped up mainframe companies and discouraged imports, affecting research directions. HCI research in Asia emphasized the consumer market. After focusing on methods to design and develop Japanese-language, individual productivity tools for the domestic market, and technology oriented toward English for export, much Japanese research turned to language-independent communication tools.[86] As illustrated by the example of CSCW discussed above, regional differences can impede comingling.

11.5   SUMMING UP

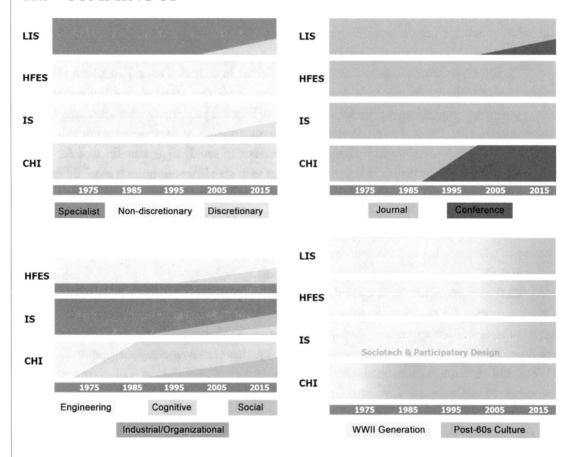

Figure 11.1: Four dimensions on which HCI fields differed, affecting priorities, methods, and ease of interaction. Clockwise from upper left, the nature of the use being supported, the venues in which results are published, influential fields of psychology, and the culture of the researchers and developers are shown.

---

[86] Tomoo Inoue, personal communication, March 2012.

Figure 11.1 is a simplified depiction of barriers to communication and collaboration. The lower left chart omits library and information science, which drew more on the humanities for insights into people, and avoided the battles among branches of psychology.

Given these impediments, should we encourage interdisciplinarity and multicultural exploration? Possibly. There is some convergence on the right of the charts. But the different fields could be headed in different directions, like matter in the universe following the Big Bang, forming different worlds of research and practice that may or may not someday discover one another and find ways to communicate.

CHAPTER 12

A New Era

12.1 SYMBIOSIS AND AI

In speculating about the years ahead, we must be mindful of Edgar Fiedler's admonition that "he who lives by the crystal ball soon learns to eat ground glass." That said, our field is unusual in having five decades of strong change that in some respects has been in one direction. As seen in Chapter 6, this led to some clear trends. In *The Innovator's Dilemma*, Clayton Christensen (1997) examines several technology disruptions such as mechanical excavators and electric cars; the one digital example is by far the strongest and most convincing. Moore's law may slow, but size, cost, and power consumption continue to decline and technology continues to spread to far reaches of the planet and engage with more aspects of our lives.

The changes that appeared approximate every decade were driven by hardware. Now a major qualitative transition is occurring at the level of software. The early visions are realized and we are edging into a new phase. Where a half century ago we had at most limited prototypes, mature products are now in routine use. Ivan Sutherland's icons, constraints, and the computer-animated movies that he predicted are here. All interactive devices incorporate the elements revealed in Engelbart's "mother of all demos." The dynamic web supports the core of Ted Nelson's hypermedia and tablets embody Alan Kay's Dynabook. Not every imagined feature exists, but all have been explored, and many that were not envisioned have appeared. Let's revisit J.C.R. Licklider's 1960 blueprint, as described in the Chapter 1 and elaborated in Chapter 5.

- Phase 1: Human-computer interaction. Create usable technology for input, display, and information organization and processing. Done! Not perfect, there is always more to do, but it is time for collective celebration.

- Phase 2: Human-computer symbiosis. People and machines work together as partners, intellectually the most creative and exciting time in history. This is not widespread, but it is appearing.

- Phase 3: Ultra-intelligent machines. Licklider was uncertain about the timeframe, but funded respected colleagues who insisted they would arrive soon.

Phase 2 did not arrive with a ribbon-cutting ceremony and fireworks. It slipped in, perhaps as we slept, when our devices, servers, and applications began working around the clock on our behalf, semi-autonomously, gathering, filtering, and organizing our information. Our history and location are used to prioritize incoming information and formulate suggestions. An automobile may warn us as we approach objects or automatically trigger a collision avoidance maneuver. Turbotax offers to take the initiative or leave it to us. It is not a balanced partnership. Our software partners can be frustratingly obtuse—and our frustration proves that we know it could be better. It will be better. We have embarked on a journey.

The concept of symbiosis elicits the contrast of technology that augments human action (tool) with technology that replaces human action (AI) that first arose in the 1960s.[87] Douglas Engelbart considered human augmentation, not automation, to be the goal. He felt the AI community obstructed him as claims that superintelligent machines would soon automate all human work got the upper hand. Engelbart's augmentation papers appeared at office automation conferences.[88]

Technology has automated assembly-line manufacturing and the work of connecting telephone calls and processing photographs, among other things. A digital partner assists in that respect it is more like a tool. It is not a person, but it is also not an alien species, arriving from outer space or brought to life with a jolt of electricity. At its best, it is a distillation of what we know about a particular task: the intent, the prerequisites, how people carry out the task, how they are motivated and demotivated, what can go wrong, and what the person or group at hand brings in the way of skills and experience. It builds on our understanding of human behavior, an understanding that is very slowly won, because we process so little of it consciously. Our autonomy need not be threatened by a good digital partner any more than by a library, which is another knowledge resource created and assembled at great effort.

However, at least two serious challenges face designers and the rest of us. Just as an individual frequently has conflicting goals, software partners face divergent interests, such as satisfying the owner and diverse users of a website. Software that knows more about all parties might better resolve the trade-offs, but it will always be a balancing act. A collision avoidance system may have to choose between a course of action with a 30% chance of injuring its driver and another with a 70% chance of injuring a pedestrian. With autonomy comes responsibility. Second, a human partner who is knowledgeable, capable, and graceful in one context is generally the same in a related context, whereas software is often totally incapable outside a narrow focus.[89] Designers must learn what people expect of partners, and we all must become familiar with non-human limitations of even the best machine partners.

Symbiosis extends interaction, but fully embracing this perspective changes how designers and developers work. Umer Farooq proposed preserving the HCI acronym, rechristening the field "human-computer integration."[90]

---

[87] Markoff's *Machines of Loving Grace* (2015) details the early history.
[88] Engelbart later chose to present his 1997 Turing Award lecture at CSCW'98.
[89] Examples are in my reflections on symbiosis near the end of Appendix A.
[90] Farooq and Grudin, 2016.

AI and HCI are both needed to get far with mixed initiative and shared control systems, and they are not inherently incompatible.[91] Visual and language processing, robotics, learning preferences, and outcome assessment will be necessary. Deeper understanding of human partnerships, cultural differences, and domain differences will help avoid the fatal limitations and awkwardness of early digital partners, such as Microsoft's Bob and Clippy. Even constrained tasks such as filing taxes, applying to college, learning to use a new game, and exploring a shopping site differ on dimensions such as the stress level of users, something a good human partner detects and responds to. AI systems must be usable and intelligent software will be more useful.

Despite the logic pointing to collaboration, there is competition for resources. During AI winters, HCI flourished. Ivan Sutherland gained access to the TX-2 when AI suffered its first setback. Major HCI laboratories formed during the AI winter of the late 1970s. HCI hiring in computer science departments swelled in the 1990s following the Strategic Computing Initiative failures. At an NSF review panel in this period comprised of half AI and half HCI researchers, an AI spokesperson insisted that 100% of the grants go to AI, "because we have mouths to feed." During AI summers, funds and students go to AI and HCI progresses slowly. For example, AI-oriented MCC attracted many HCI researchers to work on AI projects in the late 1980s.

The conflicts were sharpened by differing views of the duration of Phases 1 and 2. If the Singularity will arrive soon, long-term interface efforts are wasted: A superintelligent machine will clean up interfaces in minutes! To propose to work on HCI is to question the claims of your AI colleagues and the agencies funding them. If fully autonomous vehicles will be on the road in five years, extended research to understand *human* drivers is not useful. In contrast, if superintelligence will require 250 years and fully automated city driving will take 50, HCI projects along the way make sense. But it is difficult to get funding for 50- or 250-year projects. A promise of quick results gets attention and others then jump on the bandwagon to be competitive. Keep in mind that J.C.R. Licklider, John McCarthy, Herb Simon, Marvin Minsky, and many others predicted the singularity would arrive to end Phase 2 before 1990. A quarter century later, symbiosis has arguably just begun.

If AI and HCI researchers and developers conclude that Phase 2 will last a century or two, they can form a partnership that draws on the full range of the capabilities of each. Apart from the difficulty of resisting the temptation to promise quick results, the timing is good. AI features *that succeed* in discretionary applications and large systems will capture attention in CHI and human factors.

12.1.1 THE AI ROSE BLOOMS AGAIN

A hot AI summer was in full force by 2016. Vast sums were invested in driverless cars and trucks, with media covering every round of financing and every prototype crash. Many forecasted fully autonomous vehicles rolling off assembly lines as early as 2021. Tesla announced that the hardware

[91] Fischer and Nakakoji, 1992.

for full automation was ready to go into new cars, with software to follow.[92] On a related front, impassioned writers agonized over impending job losses through automation. An eminent panel of four U.S. economists and four technologists met in 2014. The president of Stanford Research Institute forecast that in 15 years, unemployment would be universal; a speaker from Singularity Institute predicted that it would only take five years for capable machines to take over. Their concerns echoed Marvin Minsky's 1970 lament that, "If we're lucky, they might keep us as pets." (The economists were less worried.) IBM was heavily promoting the potential of its Jeopardy-winning Watson software. "Deep learning" seized imaginations. Start-ups started up. Small AI companies and researchers were acquired or hired by Google, Apple, Facebook, and Microsoft.

After one major player makes an optimistic estimate, others follow suit. An automobile company promises fully autonomous vehicles in five years and its competitors jump into the fray. IBM proclaims that Watson will revolutionize computing and its competitors follow. In 2015, a U.S. government research laboratory manager was asked why so many HCI programs were actually AI projects not dissimilar to those that failed in the previous AI summer. The pained response was, "What if this time it's real? We can't afford to miss it."

Is it real this time? Not all HFES and CHI researchers think so. AI pioneer Roger Schank's (2015) essay arguing that IBM's Watson AI program is "a fraud" concluded, "AI winter is coming soon." A senior IBM researcher working on cognitive services wrote that he agrees with Schank. Another expert wrote in 2016 about driverless cars, "The public is constantly misled by overly enthusiastic automotive reporters who don't really understand the technological challenges on the one hand and enjoy promoting hype on the other. The question is who will take the responsibility of conveying the reality to the public? [Those in the companies] cannot really do it in a fair way, the federal government officials have their hands tied in bureaucratic knots, and academicians are silent."

Even if fully autonomous cars fail to materialize, machine vision and other technologies can improve collision avoidance and assist with parking. Machine learning succeeds in constrained tasks such as face recognition and algorithmic trading. This is progress. However, when deep learning and data science displace behavioral analysis of software use, rather than partnering with it, most progress may be postponed until after failures to meet promised delivery dates, which history suggests are likely.

Each AI summer sees discussion of the arrival of Phase 3. Today, it is more cautious than the 10- and 20-year projections that stretch back to 1960. Most singularity fans place it 30 to 40 years out, according to a nice Wikipedia entry. Informed skeptics argue that it will take at least a century.[93] AI has not tackled the small and large inefficiencies that plague daily interactions with

---

[92]   Tesla, 2016.
[93]   Allen and Greaves, 2011.

devices. Detecting problems is possible; fixing them automatically is not. (Detecting them by asking or observing users can be easier.)

Whether the result of the singularity or simply more capable machines, the issue of a coming job decimation became a major media focus in 2016. Jobs disappeared: hundreds of thousands of telephone operators, photographic film processors, travel agents, and secretaries. But overall employment continued to rise. Poverty worldwide dropped sharply. In the U.S., both total and unfilled jobs reached record highs in 2016. Productivity gains per worker accelerated from 1995 to 2005, then dropped back to historic norms.

Automation disrupts some industries, but new kinds of jobs proliferate. History may be instructive. Agriculture disrupted hunting and gathering, and the industrial revolution disrupted farming. With those disruptions, new jobs and job categories appeared, and today, entirely new job categories appear, from UX professional to microbrewery engineer.[94]

This rest of this chapter is organized as follows. Two inevitable consequences of hardware advances are explored: the "Internet of Things" and domain-specific HCI. Next, major methodological shifts toward design and metrics are considered. Then we take a final look at HCI's flashy cousin, AI. The chapter concludes by focusing on two profoundly consequential effects of people interaction with information technology: increased transparency and decreased community.

12.2   INTERNET OF THINGS AND PEOPLE

As costs decline, people find ways to make technology useful and appealing in a wider variety of tasks. The focus of youth on peer communication is driving the evolution of smartphones and social media. On streets and campuses, students seem tethered to devices. An open question is how much of our digital activity can be conducted on small displays. Less visible but perhaps more salient is the shift toward embedded technologies.

When visible, in watches, bands, helmets, and glasses, their use may be limited to specific contexts or tasks, such as fitness and entertainment. Invisible sensors that collect information will be elements of our partnership with technology. As with smartphones, the technical and behavioral infrastructure has to be worked out. Energy harvested from light, heat, and body movement will be exploited by more efficient hardware. Overcoming challenges in networking these devices and filtering, aggregating, transmitting, visualizing, or otherwise absorbing the information will be an ongoing exploration.

In the past, new users outnumbered mature users. This is always true for a new application, but not for technology use in general. As users and technologies progress along the path in Figure 12.1, the relevant priorities and methods shift.

[94]   A startling number of people are cheerfully employed training kids under 18 in every branch of athletics, music, and dance, in aca-demic tutoring, SAT training, college application writing, scholarship chasing, and so on.

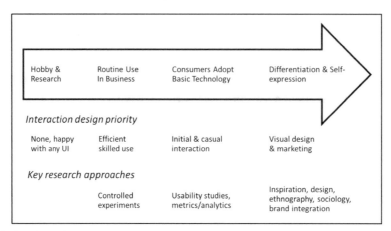

Figure 12.1: Technology, from invention to maturity.

Scientists tinkered with watches for centuries. Watch manufacturing arose in the mid-19th century when railroads needed reliable timing. After World War II, Timex benefited from a mass market for inexpensive watches. Consumers motivated by cost and efficiency were happy to have identical watches, just as early motorists were fine with identical Model T Fords. Swatch realized that people hankered for differentiation, as General Motors had when it designed cars for different uses and status claims.

When a technology reaches the final step in Figure 12.1, new challenges surface. Science strives for generality, but now differentiation is the goal. Is there room for basic science here? A technology focused on a specific domain or market segment seems applied. Could research that applies to affluent teenagers or to hospital organization be basic? Yet, careful research is required to support different groups and industries, hospitals, banks, real estate firms, department stores, firefighters, and so forth. The shift from designing for the greatest common denominator in a large market to differentiating for many smaller markets is significant, and must be carefully timed. Ford did well in the 1920s, General Motors in the 1950s. Timex thrived in the 1950s, Swatch in the 1980s. Indistinguishable PCs were fine in the 1980s; iMacs were a successful differentiation in the 2000s. HCI must reposition when a shift comes. In the past decade, differentiation led by design has pulled practitioners ahead of researchers.

12.3 DOMAIN-SPECIFIC R&D

Rapid change stresses academic disciplines and professional organizations, whose growth is outpaced by the demand to support more activities at increasingly finer granularity and technology diffusion into industries that have distinct needs. Academics, professional conferences, and journals that focus on general methods and theories address a shrinking proportion of practitioner needs for maturing technologies in established domains.

HFES and IS focus mainly on the second column of Figure 12.1. HCI research in HFES has shifted from CSTG to domain-specific work in aeronautics, aging, health (which recently became the largest technical group), and other areas. Similarly, NAS BOHSI is undermined as other NAS groups, centered on specific domains, take on human factors work that BOHSI considered within its charter. In IS as well, the uniform system management principles of the past are less applicable when unique customer and vendor relationships emerge for different businesses. General organizational theory gives way to approaches for identifying unique requirements for the business at hand, such as data analytics and market research. Initially, quantitative analytics is a general enough method for the skill to be transferable, but over time, domain-specific analytics will be needed.

The LIS field began in the second column as well, supporting a specific domain: libraries. The embrace of iSchools reflects the awareness that information is now important in all domains. Perhaps general theory and expertise will prove to be useful; perhaps information management will also become domain-dependent.

CHI began in the third column and is shading into the fourth. When use becomes routine and progress is incremental, the most useful methods are in column two and the work leaves CHI. As new domain-specific technologies required attention, CHI membership declined as "UX specialists" congregated in conferences such as UXPA and regional gatherings, including local CHI chapters. As CHI's domain-specific efforts in software development, health and fitness, and so on mature, that work could move to HFES or elsewhere.

12.4   ATTENTION TURNS TO DESIGN AND ANALYTICS

How much influence can one individual have on the direction of history? Science fiction explores this with parallel universes. The real world had natural experiments through geographic isolation. Many technologies were invented independently in different regions, but some only appeared once. The wheel was not put to use in the Western Hemisphere, so an unknown Eastern Hemisphere inventor may deserve considerable credit. The printing press and gunpowder appeared to be latent for centuries before their potential was realized. In Isaac Asimov's Foundation science fiction series, history, largely predictable through statistical methods, is temporarily disrupted by a powerful mutant.

How HCI would have progressed without the influential engineers of the 1960s we cannot know. In the 21st century, Steve Jobs arguably has had the greatest impact. In 2007, as he laid off the HCI people hired during his absence, he reportedly said, "Why do you need them? You have me." Whether the successes of the iMac, iPod, iPhone, and iPad were due to Jonathan Ive's beautiful designs or a unique genius of Jobs for ensuring that the products had a strong consumer appeal,

HCI research outside of design was not involved. The UX balance shifted to visual design. Rival technology companies followed Apple, hiring designers and in some cases dramatically reducing the ranks of other HCI professionals.

There was a logic to this: Design was suddenly largely free of memory and processing constraints. But did the pendulum swing too far away from the complexities of interaction and usability? Some design-driven products such as Microsoft's Windows 8 could have benefited from more user consideration. If post-Jobs Apple products such as its digital watch disappoint, it could swing back. Design could again be a peer of qualitative exploration, laboratory examination, and online analytics.

It requires no crystal ball to see that "big data," usually user data, will continue to ascend. It is a key hiring area in information systems, information science, and computer science. Many companies desire skilled data scientists. Endless studies mine Twitter streams and the Wikipedia revision histories. Student interns are keen to work with corporate researchers to analyze proprietary data. A/B testing revolutionized website evaluation. Customers become unpaid study participants. Software and hardware metrics routinely report use, preferably with user permission.

However, not all possible permutations and combinations can be tested. As noted in Chapter 10, A/B testing and metrics have limitations. An attractive solution is to combine quantitative and qualitative metrics to identify patterns that qualitative research explains, and iterate. But qualitative research is slow and quantitative researchers often prefer to squeeze out weak causal relationships when other methods could be more efficient. Is it more efficient to run A/B tests until something works well enough? Ants forage by dispersing workers in every direction to see which return with a prize. Ants outnumber people, so maybe they are onto something.

Quick iteration often suffices for new applications. Mature, symbiotic support may require an understanding that qualitative methods can best deliver. This may be recognized in specific vertical markets before it is realized by technology companies dazzled by the successes of Apple designs and Google algorithms.

12.5   NO PLACE TO HIDE: OUR COMPLEX STANCE ON VISIBILITY

Asynchronous access to information and awareness of distant activity have been major goals of HCI. Early achievements were greeted enthusiastically. A decade later, we found that there could be too much access and awareness: privacy and overload issues surfaced. The trend is toward greater visibility, but some prominent technologies have not thrived. Some were abandoned after showing strong efficiency benefits. It is worthwhile to understanding the dynamics.

Group decision support systems (GDSSs) were prominent in the early 1990s (Chapter 8). Despite yielding impressive benefits in extensive trials, they failed to get traction. Managers disliked ceding authority to meeting facilitators. The anonymity of contributors, designed so ideas would be judged on their merits, interfered with customary practices of assessing contributions and awarding credit.

Second Life was used at NSF in the late 2000s (Chapter 10) for 30 grant proposal panel reviews. The panels were effective and achieved tremendous savings by eliminating travel, hotel, and meal expenses. But Second Life reviews were discontinued. Program directors were not enthusiastic. They valued meeting panelists, and it was difficult to gauge the engagement of the people behind the avatars during meetings.

Desktop videoconferencing studies that targeted workplace communication unified CSCW in the late 1980s and 1990s (Chapter 8). Members of research labs set up video cameras and let colleagues view their activity. In one case, two had a continually running link for over three years.[95] Published papers describing the promise lured researchers with more powerful computers to try again. Experiments at Sun Microsystems, Fuji Xerox, and Microsoft in the mid to late-1990s were reported at CSCSW and HICSS. However, warning signs emerged in Media Richness studies. Williams (1997) found that video effectiveness was limited to meetings with sharp conflicts or native speakers of different languages. Collaboratory researchers were startled by the low use of videoconferencing.[96] Desktop videoconferencing finally became accessible via Skype and other products in the mid-2000s. It is used for informal two-party conversations. For workplace teleconferencing, document-sharing is valued and strong audio quality is necessary—and often sufficient. Video is surprisingly often not turned on, even when it would require only a button click. A high-resolution still image is often the preferred visual representative. Video reveals attire, hair, and surroundings. It reveals inattentiveness when the discussion moves outside someone's area of concern. Even with people we know, we are less animated when we interact via video than in person, whether due to the lack of arm gestures, pheromones, or something else.

Nevertheless, video is on the rise: Surveillance cameras record us constantly. The rise of "big data," primarily documenting our activity, is taken in stride. More researchers press companies to provide greater access to their servers than press them to record less. If unfortunate consequences lie ahead, will we detect them and push back?

Ordinary citizens and politicians, police and lawbreakers, peaceful protestors and violent rioters are captured digitally in unguarded moments. Context may be missing as we consume tabloid articles about the lives of celebrities and follow politicians' gaffes and mobile phone recordings of crimes. However, the greatest challenge may be subtler than partial misunderstandings, privacy incursions, or distraction. Direct viewing exposes illusions, shows us that people behave not as we thought they did or as we think they should, reveals the corners that people cut to work efficiently and live comfortably. We realize that social conventions, policies, regulations, and laws are not con-

---

[95] Dourish et al., 1996.
[96] Olson et al., 2008.

sistently followed and violations are not uniformly punished. We see our leaders' warts, which can breed cynicism and undermine support for public institutions.

Researchers who discover that their germinating ideas are less original than they thought could be less motivated to develop an idea to the point where it is new. An Amazon search for books on creativity returned over 36,000 titles. The feverish attention to creativity from magazines, schools, and NSF could flow from technology that reveals how little is new under the sun.

These issues are more pressing as machines become more capable. Consider a few examples, starting with speed limits. Technology could enforce total compliance, but, so far, speed camera placement is negotiated. Should a car with autodrive ignore speed limits that it is aware of? Who will buy a car that rigidly enforces limits or nags frequently? Political candidates can no longer tailor their message to the audience at hand, nor can managers safely reshape messages from above for their groups. Another example: some company policies are not intended to be followed literally; the message is, "Be very careful; if you take a short cut that causes damage, you are responsible." Software that enforces such policies can reduce productivity; selective enforcement when that too is visible raises questions of fairness. World religions that tolerated diverse practices in different regions encountered this issue when the web revealed the discrepancies. Robert Ellickson's (1991) *Order without Law* reveals these complexities of behavior that technology surfaces. How will individuals, families, organizations, and societies approach these challenges? Working out nuanced rules for all contingencies is impractical. Perhaps we will learn to tolerate more deviance in behavior than we have been comfortable with, as technology exposes the world as it is.

Symbiosis potentially amplifies this. One need not fully embrace the concept of a digital partner—a tool that is better fitted to our hand or eye will be more effective—but a partner who knows more about us will be more effective and less awkward. Online businesses have an incentive to retain our trust, but an inherent conflict remains: when does knowing about us help it serve us, and when does it reveal exploitable weaknesses? We may unconsciously assume that a truly helpful, friendly seeming software partner will respect boundaries as a friend would, for example, by being discrete about passing on information. But this could well be wrong; it will do whatever it is programmed to do. One can look closer to home than drones to see that boundaries are constantly tested. Who benefits from the upselling of deluxe versions or ancillary applications? Data on children's online activities that are collected to improve education could be used to enhance subtle marketing. What one side feels is well intentioned may not appear that way to another.

We decide how much about ourselves to disclose to achieve good relationships with people. The more our digital partners know about us, the better they can help us. But even with people, the boundary between assisting and exploiting is fuzzy. Being guided to an irresistible purchase can be great or harmful for different people or at different times. An online gambling opportunity that is fun for her is a dangerous addiction for him; a chocolate cake recipe that is wonderful for a party planner is unfortunate for a dieter. Whether an offer is a net positive, net negative, or neutral may never be clear. And our digital partners are intrinsically amoral: a tool that does us a great favor one moment could easily do something utterly detrimental the next, in a way that people around us rarely do. HCI is becoming more complex.

12.6   COMMUNITY AND TECHNOLOGY

Communities and technology intertwine in two ways in this book: Computer use can support or undermine communities, and professional communities study HCI. Online community efforts flourished as soon as computer-mediated communication was possible. Chapters 7 and 9 covered the immensely popular community bulletin boards of the 1980s and 1990s. As modem-based dial-in systems, their membership was mostly local and in some cases planned face-to-face get-togethers. The best-known was The WELL (Whole Earth "Lectronic Link"), started by Stewart Brand and Larry Brilliant in 1985 in the San Francisco Bay Area. Usenet newsgroups were active community forums prior to the web that by design had no geographic focus.

In 1991, the Santa Monica Public Electronic Network was launched by the municipal government with public kiosks supplementing private access.[97] The few such geography-based efforts had limited traction. Today, there are many one-way local broadcast networks for schools, homeowner associations, and the like. The extent of local online discussion groups may be understudied, as were community bulletin boards.

Geographically dispersed groups have multiplied radically. Social networks often started with local friends and colleagues and then spread to distant relatives, past schoolmates and colleagues, professional acquaintances, and others. Professional organizations create forums, as do product vendors interested in marketing or feedback. Within each category, the number and size of communities proliferates. People who initially followed one social networking site are often active on several—Facebook, LinkedIn and Twitter, Snapchat and Instagram, and others. In the field of HCI, conferences have proliferated, each striving for a community of participation.

Basic arithmetic says that our focus is moving outward. Peripheral activity and community compete for our attention. We have a fixed number of hours available for social interaction. Social lives at college once centered in dormitories; now students carry their dispersed high school cohort in their pocket. Pedestrians glued to phones and employees riding elevators with phones in front of them lose touch with neighborhoods and colleagues. "We've lived in our neighborhood thirteen years, and I've had only four interactions with other residents," said an acquaintance. For many professionals, including athletes, academics and others, the ease of finding new jobs leads to mobility and shallower roots. When past colleagues are keystrokes away, the cognitive and social drive to form new close ties diminishes. The significance of the decline in community cannot be exaggerated.

[97] McKeown, 1991.

For millions of years, our ancestors lived in extended families, tribes, bands, or communities. Most primates survive by being a member of a troop, cooperating to gather food and repel predators. Our tribal past lives in us. Soldiers report fighting for comrades more than for country. Yet our species is also driven to disperse, in groups: Over a few dozen millennia we spread across the globe, occupying niches from arctic ice to equatorial jungle, arid desert to oxygen-thin mountaintop, continually inventing new cultures and languages.

Transportation and communication technologies extended our geographic range. Now a global digital technology enables us to interact directly with any one of billions of people. Theoretically, it could strengthen close communities or foster more distributed social networks. The latter appears to be more prevalent.

In pre-internet university departments, faculty were motivated to acquaint colleagues and students with their work to get quality feedback. Now, faculty can get such feedback from distant others. Secretaries were once the social glue, remembering birthdays and circulating get well cards. Few remain. Students are told to expect to change jobs very frequently, the subtext being work colleagues will rarely constitute a close community worth investing in.

Some fear that online, we will form stratified echo-chambers of the like-minded, appealing to our tribal nature. Others argue that geography-based echo chambers were once the norm, and online proximity will broaden people, appealing to our urge to disperse. The places to look most closely are among the young and educated, and our own professional communities as early adopters.

Changes in HCI fields may be informative. A new HCI discipline arose every decade, with trade shows, conferences, professional organizations, journals, and disciplines. Some aligned with technologies that reached dizzying heights and fell. Successes, such as SIGGRAPH and SIGCHI in ACM, spawned dozens of smaller conferences. We reviewed forces that inhibited collaboration across fields. The growing volume of information to be mastered within a single field may also promote specialization, as suggested by Herb Simon (1981) in *Sciences of the Artificial*.

Counterintuitively, as users went from thousands to billions, the size of the major HCI communities plateaued or declined. HFES membership plateaued in the 1990s, with CSTG declining as other groups took on human factors efforts. In IS departments, HCI rose in prominence in the early 2000s and then leveled. In some schools of management, IS merged with other groups. Institutional membership in the iSchool consortium is expanding, but iConference attendance peaked in 2011 and ASIS&T membership is declining. ACM SIGIR conference paper submissions declined after 2012 as several specialized information retrieval conferences formed. Membership in ACM SIGs has declined steadily since the early 1990s, as seen in Table 12.1. SIGGRAPH has a popular trade show and CHI membership is effectively free for conference attendees. New SIGs have started, but as is evident from the bottom row, they have not had much draw.

Table 12.1: Ten ACM Special Interest Groups had 3,000 members in 1990. (As of 2016, two of 37 SIGs have 2,000 members.)

SIG	1990	1995	2000	2005	2010	2015
Plan	12,335	6,574	4,362	2,765	2,323	1,965
Graph	11,811	6,986	6,298	7,902	7,216	7,299
Soft	10,824	6,389	3,313	2,916	2,489	2,176
AI	8,955	3,707	1,917	1,559	1,123	900
OPS	6,801	4,154	2,356	1,793	1,828	1,528
CHI	5,023	5,186	4,950	4,409	4,415	3,691
ARCH	4,685	3,035	1,730	1,454	1,384	1,348
ADA	4,089	1,867	685	391	292	209
MOD	3,952	2,959	2,941	2,317	1,922	1,490
MIS	3,082	1,442	755	636	497	293
(All 30+)	103,489	65,698	47,042	44,600	41,008	28,195

CHI's loss of practitioner membership was offset by its status as the primary quality venue for a growing academic field. CHI conference submissions increased; a 2001 peak conference attendance was not surpassed until CHI 2013 was held in Paris. "Spin-off" conferences formed and sub-disciplines were established, but the community that once saw the parent conference as a must-attend event is gone.

12.6.1 DISPERSION

Rising rejection rates in ACM conferences correlate with spin-off conference formation and the departure of practitioners. Rejection encourages researchers and practitioners to submit work to smaller, more focused conferences, which then absorb finite travel budgets and time. For example, CHI reduced its acceptance rate to 15%–16% in 2002 through 2004. It raised it to 25% in 2005 after submissions dropped dramatically, but as CHI attendance dropped, a host of smaller conferences attracted CHI authors. The DUX series began in 2003; UPA initiated a peer-reviewed paper track; HCI minitracks at HICSS thrived; more CHI authors attended HCI International; and regional HCI conferences did well. Citations rose sharply in 2003 for ACM GROUP, where many CHI and CSCW rejects are resubmitted, and GROUP 2005 citations set an all-time high, not equaled before or after.

Geographic dispersal of our species led to linguistic and cultural innovation; a multitude of smaller, focused conferences can accelerate growth by convening all researchers and developers intensely interested in a topic. Energetic younger participants acquire organizational and leadership skills. But there are drawbacks. Attention, organizational energy, travel funds, and the literature are fragmented. Wheels are reinvented in different venues. Large annual meetings of neuroscientists, anthropologists, and other academic and professional disciplines establish an overarching sense of identity. These conferences are for community-building, leaving quality to journals. From 1951 to 1987, the National Computer Conference (under different names prior to 1973) served this purpose for all of computer science.

The information field is a petri dish for examining these processes. It has evolved rapidly since digital information became a force around 2000. Quoted above, Wersig suggested that information schools could act as magnets sucking content from other fields and restructuring it. They got off to a good start, attracting faculty from library science, computer science, management science, information science, and other disciplines. Glushko (2016) offers a concerted effort to provide the restructuring. The eagerness of scores of institutions to pay dues and join is promising. However, the iConference and ASIS&T are following a path that fragmented computer science, lowering acceptance rates to around 30%. Attendance has stalled. MIT, Stanford, and all save one Ivy League school have not joined. Many information school faculty recruit from their discipline of origin, which signals to graduate students that perhaps they, too, should have another anchor. One dean described the iCaucus and ASIS&T as "two large ships adrift, moving wherever the currents take them, possibly closer together, possibly apart." A 30% acceptance quota pushes some sailors to abandon ship and discourages sailors from other disciplinary vessels from boarding.

Consolidation of information studies is threatened by the rapid rise of specialized programs—biomedical informatics, social informatics, community informatics, disaster informatics, and information and communication technology for development (ICT4D). Some event could crystallize this super-saturating solution, but more likely the sub-disciplines will follow individual trajectories away from the center.

The drift of human factors efforts from CSTG to other technical groups and from IS to other management departments also represents diffusion of HCI work. This has been a pattern in computer science. NCC disbanded when specialized conferences such as SIGGRAPH and SIGCHI attracted researchers. Participation in the SIGSOC declined in 1981, leading to the transition to SIGCHI, because it drew people from several behavioral science disciplines interested in software tools for managing laboratories, running experiments, and analyzing data; interest in that rose to the extent that meetings were convened at the annual conference of each social science. People saw no need for the joint meeting, which survived by shifting from technology support for social science to social science support for technology.

The occupation of new ecological niches could in theory be followed by an inward focus on building communities, but the large conferences weigh so heavily in academic appointments and promotions that the spin-offs are useful stepping stones but not home to many. A greater number of weak ties prevail, with few strong ties. The cost is a loss of the sense of security that our primate genes seek through membership in an extended family. Short-term benefits may exact long-term costs, but for now, if our field of early adopters is the canary in a technology mine, most of us are chirping happily, with only occasional distressed tweets.

CHAPTER 13

Conclusion: Ubiquitous Human-Computer Interaction

In human factors and information systems, HCI research was the province of one group but is now diffused throughout most working groups and academic sub-disciplines. In computer science, research and practice are now widely dispersed and heading in different directions. Information has given rise to myriad informatics disciplines before ever having fully coalesced. With the four fields that realized the early goals now losing control over the paths we are racing along, how does our understanding of their history help? Why look back?

There are reasons. As technologies appear and age, approaches should be carefully matched to contexts. Integration, or software as partner, may become the primary HCI focus, but interaction is the foundation on which it is built. Understanding the trajectories and dynamics of professional organizations can help us allocate our time and effort.

The core skill of understanding both people and technological possibilities is more valuable than ever as we seek finer-grained support of all aspects of life. It will take substantial work to make technology less awkward and frustrating. Human nature is less malleable than technology, so the effort is necessary. We gain insight into our often unconscious processes and preferences from the trials, errors, and successes of our predecessors.

Successive hardware platforms supported similar activities in new ways; at times, history does repeat itself. In the 1980s, email was embraced by students as an informal communication medium and regarded with suspicion by organizations. A decade later, when email had become a widely accepted medium for formal communication, instant messaging and texting were embraced by students for informal exchanges and regarded with suspicion by organizations. Another decade passed; texting was used everywhere, and social networking was embraced by students and initially regarded with suspicion by many organizations. Those who understood the past more quickly adopted nuanced policies and practices.

In the 1980s, as mainframe sales plateaued, government and industry invested heavily in "high performance" parallel supercomputers. Decomposing computational problems into parallel processes whose output could be reassembled efficiently proved to be difficult. Meanwhile, companies bought thousands of PCs, weak devices with little memory that were difficult to network. Almost unnoticed, PCs grew stronger, increased storage, and got networked, and the internet and the web surfaced. A decade later, as PC sales plateaued, government and industry invested large sums in "high performance" multicore PCs and parallel computing—and rediscovered the limitations.

Meanwhile, people bought millions of mobile phones, weak devices that over time grew smarter, increased storage, and got networked, a major phenomenon that took many by surprise.

Some predicted that mainframes would disappear, and later that personal computers would disappear. But enterprises kept one mainframe and households will likely find uses for a desktop computer. Just as a pilot won't monitor and manage the flight deck with a phone or other small display, large displays can help us monitor and work with a growing mountain of information of personal interest. As phone ownership approaches saturation, embedded systems creep up: sensors and effectors, weak devices with little memory and hard to network. This time, discussion of the "Internet of Things" has preceded its arrival. We may or may not be fully prepared when it comes.

Norman (1988) wrote of "the invisible computer of the future," elaborated in Weiser's (1991) concept of ubiquitous computing. Like motors, computers would be present everywhere and visible nowhere. Clocks, refrigerators, and cars have motors, but no one studies human-motor interaction. Years later, at the height of the Y2K crisis and the internet bubble, computers were more visible than ever. But today, we don't call smartphones or e-readers "computers." Invisible, ubiquitous computing is here, and so is invisible, ubiquitous HCI.

HCI is now an all but invisible presence in product design, accessibility, sustainability, health care, supporting the aged, and other tasks. Just as builders are familiar with motors, developers embed interfaces without attending conferences or reading papers. J. C. R. Licklider anticipated that the most intellectually creative and exciting period in our history would be ushered in by the shift from digital tools to digital partners that smoothly take the initiative at times. To realize the potential will require researchers and developers who, like Licklider, see technological possibilities and understand the human partner. It will require familiar skills and others yet to be discovered.

The first generation of computer designers, users, and researchers grew up without computers. Many of the second generation first encountered computers as university students and then entered workplaces and changed them. The current generation grew up using computers, game consoles, and mobile phones. In primary schools, they searched, browsed, assessed, and synthesized. Now they message and photo-share on smartphones, author with multimedia, and embrace social networking. They have different takes on privacy and multitasking. Many absorbed a design aesthetic. They are entering workplaces, and will change everything once again.

Appendix A: Personal Observations

Personal experiences can add texture and a sense of the human impact of events that otherwise are abstract. For many years, I followed a not unusual path. I worked as a computer programmer, studied cognitive psychology, spent time as an HCI professional in industry, and became a researcher working in academia and industry research. My interest in history arose from the feeling of being swept along by invisible forces, often to unplanned destinations. My first effort at describing these forces was the paper "The Computer Reaches Out." I saw, and continue to see, computers evolving and slowly engaging with the world in ways that we, their developers and users, had enabled but not entirely foreseen. What follows is a chronologically ordered account of experiences that illustrate or inform aspects of this history.

1970: A Change in Plans

As a student, I was awed by a *Life* magazine article that quoted experts who agreed that computers with super-human intelligence would arrive very soon. The Cold War and Vietnam War were raging, but if we survived a few years, machines would do what needed to be done! We could focus on what we enjoy, not on what we had thought could be useful. I shifted from physics and politics to mathematics and literature.

1973: Three Professions

Looking for work, I found three computer job categories in *Boston Globe* classified ads: (1) operators, (2) programmers, and (3) systems analysts. Not qualified to be a highly paid analyst, I considered low-paid, hands-on operator jobs but landed a programming job with a small electronics company, Wang Laboratories. For two years, I never saw the computer that my programs ran on. I flowcharted on paper and coded on coding sheets that a secretary sent to be punched and verified. A van took the stack of cards to a computer center 20 miles away, and later that day or the next morning I got back the cards and a printout. It might say something like "Error in Line 20." I learned not to make many errors.

1975: A Cadre of Discretionary Hand-On Users

Wang acquired a few teletype terminals with access to the WYLBUR line editor that had been developed at the Stanford Linear Accelerator. I was one of the programmers who chose to abandon paper and become a hands-on computer user.

1983: Chilly Reception for a Paper on Discretion in Use

After leaving the computer field to get a Ph.D. in cognitive psychology, I took up HCI work while a postdoc at the MRC Applied Psychology Unit in Great Britain. Allan MacLean and I found that some people chose a slower interface for aesthetic or other reasons, even when familiar with a more efficient alternative.[98] A senior colleague asked us not to publish our work. He was part of a large effort to improve expert efficiency through cognitive modeling and thought that demonstrating that greater efficiency could be undesirable would be a distraction: "Sometimes the larger enterprise is more important than a small study."

1984: Encountering Moore's Law, Information Systems, Human Factors, and Design

I returned to Wang, by then a leading minicomputer company. Moore's law had changed the industry. Hardware was ordered from catalogs. The reduced cost of memory changed the priorities and programming skills for software design and development. Another cognitive psychologist in a marketing research group, Susan Ehrlich, introduced me to the IS literature. I was attending local chapter meetings of both HFS and SIGCHI. In a gesture to counter CHI antipathy toward human factors, I called myself a human factors engineer. One morning, I took the commuter train to Cambridge to see the newly released Macintosh. Realizing that few software engineers had visual design skills, I encouraged Wang's industrial designers of "boxes" (hardware) to look into software interface design, and one did.

1984: Discovering Ethnography

I read a paper[99] by Lucy Suchman that included a transcript of a rational purchasing process breaking down and requiring elaborate exception-handling. Surely the example was an aberration! I asked colleagues in the purchasing department how it worked, and they described the rational process. "And that is how it works," I said. Pause. "That is how it is *supposed* to work. It almost never does." I realized that ethnographers are trained *not* to highlight unusual cases, unlike software engineers who seek unusual cases that could break a system.

1985: The GUI Shock

In the early 1980s, Phil Barnard and I were among the cognitive psychologists working on command naming, an important topic in the era of command-line interfaces. Our ambition was to develop a comprehensive theoretical foundation for HCI. But the success of the Mac in 1985 curtailed interest in command names. No one would build on our past work—a depressing thought—

---

[98]   Grudin and MacLean, 1984.
[99]   Suchman, 1983.

and it dashed our hope for a comprehensive theoretical foundation for HCI. Time to choose: were we cognitive psychologists or computer professionals? Phil remained a psychologist.

1986: Beyond "The User": Groups and Organizations

I joined the industry research consortium MCC. Between jobs, I worked on papers that addressed major challenges that I had encountered in product development. (i) Products and features that we built to support groups were not used. Why was group support so challenging? (ii) Organizational structures and software development processes were painfully unsuited to developing interactive software. What could be done? Addressing these challenges was the goal of much of my subsequent work.

1988: Discovering Organizational Science

We formed an organizational science discussion group to read classic papers. The overwhelming favorite, a densely detailed exposition, described organizations as having five parts, its nature decided by which one dominates.[100] Later I realized we had liked it because MCC was a perfect experimental test: about 500 people assembled quickly with few clear goals. Each of the five parts struggled to assert itself and shape the organization.

1989: Development Contexts: A Major Differentiator

I began a two-year stay at Aarhus University. I was intrigued by the "participatory design" approach to development, which seemed sensible yet was alien to the "waterfall model" dominant in the U.S. Denmark had little commercial software development and I realized that differences in the contexts of (i) interactive commercial applications, (ii) in-house software, and (iii) contracted systems shape the research and development practices relevant to the CHI, IS, and software engineering fields, respectively. Sorting this out led to my first library research for historical purposes.[101] Perusing long-forgotten journals and magazines in dusty library corridors felt like wandering through an archaeological site. Understanding articles from these different fields surfaced the significant challenge discussed next.

1990: Just Words? Terminology Can Matter

A premonition had arisen in 1987. Susan Ehrlich titled a paper "Successful Implementation of Office Communication Systems." To me, implementation was a synonym for coding or development. With her IS background, implementation meant introduction into organizations. Sure enough, the

---

[100]   Mintzberg, 1984.
[101]   Grudin, 1991.

ACM editor asked her to change "implementation" to "adoption." What she called systems, I called applications. Language, usually an ally, got in the way.

My description of an HCI course at Aarhus University in 1990 featured "user-interface evaluation." My new colleagues seemed embarrassed. Weeks later, a book written by one of them was published (Bødker, 1990). Its first sentence quoted Alan Newell and Stu Card: "Design is where the action is, not evaluation." My turn to be embarrassed! In their in-house development world, projects could take ten years. Design was the first phase, and evaluation had a negative stigma, coming at the end when only cosmetic changes were possible. In the world of commercial products, evaluation of previous versions, competitive products, and prototypes was a *good* thing. Evaluation drew on an experimental psychologist's skill set and was central to iterative design.

Later in 1990, I was invited to be on a panel on "task analysis" at a European conference. In CHI, a (cognitive) task analysis comprised breaking a simple task into components; for example, is "move text" thought of as "select-delete-paste" or as "select-move-place"? To my dismay, the IS-oriented panel had a different definition: an organizational task analysis in which tasks are components of a broad work process. Some Europeans unfamiliar with my consumer-oriented context felt that it was disgraceful for us to claim that our simple analyses merited a claim that we were conducting task analysis.

En route to giving a job talk at UC Irvine, I was invited to speak at the UCLA Anderson School of Management. It ended badly when the department head asked a question about users. It seemed incoherent, so I replied cautiously. He rephrased the question. I rephrased my response. He started again and then stopped and shrugged as if to say, "This fellow is hopeless." At a conference a few months later, he seemed astonished upon hearing that his Irvine friends were hiring me. Later, I discovered the basis of our failure to communicate. We attached different meanings to the word "users." To me, it meant hands-on computer users. His question had assumed the *important* users to IS: managers who specified database jobs and read reports but were not hands-on. To me, use was by definition hands-on, and his question had not made sense.

In *Moby Dick*, Ahab nails a gold doubloon to the mast, and each sailor looking at it sees something different. So it was with *user*. From a CHI perspective, the IS "user" was a *customer*. Consultants called them *clients*. In IS, a hands-on user was an *end-user*. In LIS, information professionals were users and the public were end-users. In CHI parlance in 1990, end-user and user were one and the same—a person who entered data or commands and used the output. Human factors used *operator*, which CHI considered demeaning. In software engineering, "user" usually denoted a *tool user*, a software engineer.

I had considered words to be a necessary but uninteresting medium for conveying meaning, but these experiences led me to write an essay on unintended consequences of language.[102]

[102] Grudin, 1993.

1990: The Cost of Digital Memory

When in the U.S., I picked up a $1,000 200M hard drive for a colleague. They cost more in Denmark. This memory is a reminder of the constraints we faced when digital storage was not only not free, it was damned expensive.

1998: Usability Peaks?

Through the 1990s, when I used Unix boxes and Macs, my students evaluated products and noticed marked improvement over time in the usability of Microsoft apps. When I joined Microsoft in 1998, I found an astonishingly strong applied HCI effort. This effort weakened as visual design and testing through rapid iteration based on user feedback came to dominate.

2006: Domain-Specific HCI

At NordiCHI 2006, I was surprised by papers that described fundamental research undertaken in specific domains: telecom, medicine, financial, or others. Others present also felt it was a turning point.[103] The shift to domain-specific work continued, but much of it remains outside the research literature.

Reflections on Bridging Efforts

I joined efforts to build bridges from CHI to human factors, OIS, information systems (twice), design, and information science, all unsuccessful. I experienced generational and behavioral versus cognitive psychology divides. Many people joined MCC to avoid "Star Wars" military projects. I went through CHI's shift from journals to conferences and from hypothesis-driven experimentation to qualitative approaches. Averse to the use of male generics, I avoided the *International Journal of Man-Machine Studies* and turned down an invitation to a "man-machine" interaction event.

Some differences faded, others persisted. Writing a chapter for an IS-oriented book, a coauthor and I wrangled painfully with the editor over terminology.[104] Reviewing IS models of white-collar employees, I twice came up blank when searching for references to the TAM model. I was mystified. Eventually I saw the problem: TAM stands for "Technology Acceptance Model," but I was typing "Technology Adoption Model.'"

Reflections on Symbiosis

In the 1980s, I saw negative consequences of assuming that a digital partner that is effective on one task will, like a human partner, be more broadly competent. An expert system that was good

[103]   Grudin, 2007.
[104]   Palen and Grudin, 2002.

at diagnosing the relative likelihood of two types of ulcer infuriated physicians when it did not warn them that a patient could have stomach cancer, as a human ulcer specialist would have. Later, I demoed a natural language interface to a database application. It had a nice menu-driven help system, but because it could not respond to natural language help queries, people dismissed it. (It had other problems due to the nature of English, such as scoping ambiguity.)

Predictions

Can understanding forces from the past that shaped the present improve our odds of anticipating or reacting with agility to future events? Will it identify the efforts that are likely to prove futile through underestimation of either the immovable object that is human nature or the irresistible force of technology change? My first effort was fine apart from missing the commercialization of the internet on the horizon.[105] Had I thought more about history, the commercialization of software and then "look and feel," I might have seen that coming. Look ahead, but keep an eye on what is happening around you!

---

[105]   Grudin, 1990a.

Appendix B: A Toolkit for Writing a Conceptual History

I identified when empirical studies of programming, technology acceptance, information visualization, media richness, crowdsourcing, information retrieval, crisis informatics, assistive technologies, and other aspects of human computer interaction were studied, but not how they matured. It would be wonderful to read about the evolution of techniques, models, and theories. Blackwell's (2006) history of metaphor use in design is an example.

This appendix provides starting points for assembling such histories. Books that are out of print can be found in libraries and often inexpensively purchased online—the ready availability of used books online helped me immensely in writing this book.

The HCI Bibliography, launched by Gary Perlman over 25 years ago, is a searchable index of over 125,000 articles with abstracts. It includes a section on HCI history resources. (See http://hcibib.org.) Edited handbooks provide time-anchored snapshots taken by subject matter experts who strive to present comprehensive accounts of work in an area. Five handbooks titled HCI were published over 25 years:

1988: *Handbook of Human-Computer Interaction*. Helander, M. G. (Ed.) North-Holland.

1991: *Handbook of Human-Computer Interaction*, 2nd edition. Helander, M.G., Landauer, T.K., and Prabhu, P.V. (Eds.) North-Holland.

2003: *The Human-Computer Interaction Handbook: Fundamentals, Evolving Technologies, and Emerging Applications*. Jacko, J. A. and Sears, A. (Eds.) Erlbaum.

2007: *The Human-Computer Interaction Handbook: Fundamentals, Evolving Technologies, and Emerging Applications* (2nd edition). Sears, A. and Jacko, J.A. (Eds.) CRC Press.

2012: *The Human-Computer Interaction Handbook: Fundamentals, Evolving Technologies, and Emerging Applications* (3rd edition). Jacko, J.A. (Ed.) CRC Press.

These are weighted toward computer science/CHI research but have chapters on human factors, decision support, information retrieval, and other topics, written and reviewed by people from the other disciplines. The declining emphasis on journal publication in computer science makes these especially valuable as sources, as does the thin coverage of HCI work in journals such as *Computing Surveys*.

Several edited collections of previously published articles, generally accompanied by framing essays that identify additional sources, were highly influential and reveal prominent contemporary issues:

1987: *Readings in HCI: A Multidisciplinary Approach*. Baecker, R. and Buxton. Morgan Kaufmann.

1995: *Readings in HCI: Toward the Year 2000*. Baecker, R., Grudin, J., Buxton, W. and Greenberg, S. Morgan Kaufmann.

1988: *Computer Supported Cooperative Work*. Greif, I. Morgan Kaufmann.

1993: *Readings in Groupware and Computer-Supported Cooperative Work: Assisting Human-Human Collaboration*. Baecker, R. Morgan Kaufmann.

1999: *Readings in Information Visualization: Using Vision to Think*. Card, S.K., Mackinlay, J.D. and Shneiderman, B. Morgan Kaufmann.

Four editions of a human factors handbook were edited by Gavriel Salvendy and published by Wiley:

1987: *Handbook of Human Factors*.

1997: *Handbook of Human Factors and Ergonomics* (2nd edition).

2006: *Handbook of Human Factors and Ergonomics* (3rd edition).

2012: *Handbook of Human Factors and Ergonomics* (4th edition).

They are not confined to HCI, but the HCI focus grows; in 2012, an HCI section comprises nine chapters and other chapters cover aspects of HCI. For example, human performance modeling, which began in CHI and was represented in early CHI-oriented handbooks, has a six-chapter section in the human factors handbook and only one chapter in the most recent HCI handbook.

The journal *Human Factors* publishes survey articles. Two books that reward reading are the collection of HCI papers from the early human factors literature and the human factors history monograph cited earlier:

1995: *Human Factors Perspectives on Human-Computer Interaction*. Perlman, G., Green, G.K. and Wogalter, M.S. (Eds.) Human Factors and Ergonomics Society.

1999: *The History of Human Factors and Ergonomics*. Meister, D. Erlbaum.

Information systems handbooks:

2003: *The Handbook of Information Systems Research*. Whitman, M. and Woszczynski, A. (Eds.) Idea Group.

2011: *The Oxford Handbook of Management Information Systems: Critical Perspectives and New Directions*. Galliers, R.D. and Currie, W.L. (Eds.) Oxford University Press.

These include chapters by leading researchers on topics of general interest, such as virtual work, qualitative methods and quantitative methods.

Edited collections:

2006: *Human-Computer Interaction and Management Information Systems: Foundations*. Zhang, P. and Galletta, D.F. Routledge.

2014: *Human-Computer Interaction and Management Information Systems: Applications*. Galletta, D.F. and Zhang, P. Routledge.

With IS's strong journal-orientation and premium on top journals and its flagship conference, *Management Information Systems Quarterly* (MISQ), Information Systems Research (ISR), and the proceedings of the International Conference on Information Systems (ICIS) are good sources.

In information science, the *Annual Review of Information Science and Technology* (*ARIST*), initiated by the ADI and NSF and published on behalf of ADI/ASIS/ASIS&T from 1966 to 2011, can be seen as a handbook released in stages. Wiley Online sells the 2002–2011 volumes as of this writing.[106] Other years can be found in libraries and online as used books. The information field has a wealth of written histories. Few are focused on the digital era, but they provide insight into the field's embrace of technology. Several contributions to *Historical Studies in Information Science*, edited by Hahn and Buckland, have been cited in this book. Burke's *Information and Secrecy* reviews the field through the late 19th and early 20th centuries. Others are:

Aspray, W. The many histories of information. *Information & Culture*, 50, 1, 1-23. (2015).

Aspray, W. The history of information science and the other traditional information domains: Models for future research. *Libraries and the Cultural Record*, 46, 2, 230–248. (2011).

Black, A. Information history. *Annual Review of Information Science and Technology*, 40, 441–473. (2006).

Cortada, J. *All the Facts: A History of Information in the United States Since 1870*. Oxford University Press. (2016).

The most intensely documented period stretches from the mid-1970s to the mid-1980s in the San Francisco Bay Area, when personal computing made HCI salient to many. The books below reveal the range of topics and perspectives. Lengthy articles also appeared in *Wired* and other magazines.

Hackers: Heroes of the Computer Revolution. Levy, S. (1984). Anchor Press/Doubleday.

Insanely Great: The Life and Times of Macintosh, the Computer that Changed Everything. Levy, S. (1994). Viking.

---

[106]   http://onlinelibrary.wiley.com/journal/10.1002/(ISSN)1550-8382

Dealers of Lightning: Xerox PARC and the Dawn of the Computer Age. Hiltzik, M. A. (1999). HarperCollins.

Fumbling the Future: How Xerox Invented, Then Ignored, the First Personal Computer. Smith, D.K. and Alexander, R.C. (1999). iUniverse.com.

Bootstrapping: Douglas Engelbart, Coevolution, and the Origins of Personal Computing. Bardini, T. (2000). Stanford University.

Revolution in the Valley: The Insanely Great Story of How the Mac Was Made. Hertzfeld, A. (2005). O'Reilly Media.

What the Dormouse Said: How the 60s Counter-Culture Shaped the Personal Computer. Markoff, J. (2005). Viking.

Designing Interactions. Moggridge, B. (2007). MIT Press.

How a Group of Hackers, Geniuses, and Geeks Created the Digital Revolution. Isaacson, W. (2014). Simon and Schuster.

Steve Jobs. Isaacson, W. (2015). Simon and Schuster.

Machines of Loving Grace: The Quest for Common Ground between Humans and Robots. Markoff, J. (2015). HarperCollins.

The fragmentation of our literature together with the accelerating pace of change, discussed in the text, complicates historical research. I could not locate comprehensive surveys on topics that have extensive research literatures, such as Wikipedia authorship or Twitter feed analysis, crowdsourcing, assistive and sustainable technologies, and crisis informatics. As we move from print to streaming media, video snapshots may become a primary focus. For example, *Frontiers in Crisis Informatics* (Palen, 2015) is an hour-long analysis that covers more work than is available in written accounts.

Finally, while this book supersedes my past general articles and chapters on HCI history, additional details appear in the history column that I wrote and edited for *ACM Interactions* magazine from 2003 to 2011.[107]

---

[107] http://jonathangrudin.com/Tool_to_Partner.

Glossary

ACM	Association for Computing Machinery
ADI	American Documentation Institute (predecessor of ASIS, ASIS&T)
AFIPS	American Federation of Information Processing Societies
AI	Artificial Intelligence
AIGA	American Institute of Graphic Arts
AIS	Association for Information Systems
ALISE	Association for Library and Information Science Education
ANSI	American National Standards Institute
AOL	America OnLine (original name)
ARIST	Annual Review of Information Science and Technology
ARPA	Advanced Research Projects Agency (sometimes DARPA)
ARPANET	ARPA NETwork, first Internet predecessor
ASIS	American Society for Information Science (became ASIS&T)
ASIS&T	Association for Information Science and Technology (previously ASIS, ADI)
AT&T	American Telephone and Telegraph
BBN	Bolt, Beranek and Newman
BBS	Bulletin Board System
BIT	Behaviour and Information Technology
BITNET	Because It's There NETwork, or Because It's Time NETwork
BOHSI	(National Academy of Sciences) Board on Human-Systems Integration
CAD	Computer-Assisted Design
CD	Compact Disk
CeBIT	Centrum für Büroautomation, Informationstechnologie und Telekommunikation
CEDM	(HFES) Cognitive Engineering and Decision-Making (technical group)
CES	Consumer Electronics Show
CHI	Computer-Human Interaction (field and annual conference)
CMC	Computer-Mediated Communication
CMU	Carnegie Mellon University
COBOL	COmmon Business-Oriented Language
COHSI	(NAS) Committee on Human Systems Integration (became BOHSI)

COIS	Conference on Office Information Systems
COMDEX	Computer Dealers' Exhibition
COOCS	Conference on Office Communication System
CRT	Cathode Ray Tube
CSNET	Computer Science NETwork (preceded Internet)
CSTG	Computer Science Technical Group of HFES
CSCW	Computer Supported Cooperative Work (field and conference series)
CYC	EnCYClopedia (later Cyc)
DARPA	Defense + ARPA (used in some years)
DEC	Digital Equipment Corporation
DIS	Design of Information Systems (conference series)
DP	Data Processing (preceded MIS, IS)
DUX	Design of User EXperience (conference series)
DVD	Digital Versatile Disk
ECSCW	European CSCW (field and conference series)
EMACS	E MACroS (based on E, a text editor)
ENIAC	Electronic Numerical Integrator And Computer
ESPRIT	European Strategic Program on Research in Information Technology
EuroPARC	European branch of (Xerox) PARC
FBI	Federal Bureau of Investigation
FORTRAN	FORmula TRANslation
GDSS	Group Decision Support System
GLS	Graduate Library School (University of Chicago)
GMD	Gesellschaft für Mathematik und Datenverarbeitung
GOMS	Goals, Operators, Methods, and Selection (rules)
GUI	Graphical User Interface
HCI	Human Computer Interaction (used broadly here; also a journal)
HCIL	HCI Laboratory (University of Maryland)
HF&E	Human Factors and Ergonomics
HFES	Human Factors and Ergonomics Society (succeeded HFS in 1992)
HFS	Human Factors Society (became HFES)
HICSS	Hawaii International Conference on System Sciences
HP	Hewlett Packard
HPM	(HFES) Human Performance Modeling (technical group)
HTML	HyperText Markup Language

HTTP	HyperText Transfer Protocol
HUSAT	HUman Sciences and Advanced Technology
IBM	International Business Machines
ICIS	International Conference on Information Systems
ICOT	Institute for New Generation Computer Technology
ICQ	"I seek you"
ICT4D	Information and Communication Technologies for Development
IEEE	Institute of Electrical and Electronics Engineers
IJHCS	International Journal of Human-Computer Studies (was IJMMS)
IJMMS	International Journal of Man-Machine Studies (became IJHCS in 1994)
INTERACT	International Conference onHuman-Computer Interaction
INTREX	INformation TRansfer EXperiments
IPTO	(ARPA) Information Processing Techniques Office
IS	Information Systems (previously DP, MIS)
iSchool	Information School (post-secondary)
ISR	Information Systems Research
IT	Information Technology
IT pro	IT professional
JOSS	JOHNNIAC Open Shop System
Linux	LINUs (Torvalds) + uniX
LIS	Library and Information Science
Lisp	LISt Processing
LSI	Large Scale Integration
MB	MegaByte
MCC	Microelectronics and Computer Technology Corporation
MEMEX	Often memex or Memex; probably MEMory indEX.
MIS	Management Information Systems (succeeded DP)
MISQ	MIS Quarterly
MIT	Massachusetts Institute of Technology
MOOC	Massive Open Online Course
NAS	(U.S.) National Academy of Sciences
NASA	(U.S.) National Aeronautics and Space Administration
NASDAQ	National Association of Securities Dealers Automated Quotations
NATO	North Atlantic Treaty Organization
NIH	(U.S.) National Institutes of Health

NCC	National Computer Conference
NLS	oNLine System
NSF	National Science Foundation
NSFNET	NSF Network (succeeded ARPANET and preceeded Internet)
OA	Office Automation
OIS	Office Information System
PARC	(Xerox) Palo Alto Research Center
PC	Personal Computer
PEN	(Santa Monica) Public Electronic Network
PDP	(Digital Equipment Corporation) Programmed Data Processor
PLATO	Programmed Logic for Automatic Teaching Operations
RAM	Random Access Memory
R&D	Research and Development
SDI	Strategic Defense Initiative
SGI	Silicon Graphics, Inc.
SIGAI, SIGART	(ACM) Special Interest Group on Artificial Intelligence
SIGCHI	(ACM) Special Interest Group on Computer-Human Interaction
SIGGRAPH	(ACM) Special Interest Group on Graphics
SIGHCI	(AIS) Special Interest Group on Human-Computer Interaction
SIGIR	(ACM) Special Interest Group on Information Retrieval
SIGOA	(ACM) Special Interest Group on Office Automation
SIGSOC	(ACM) Special Interest Group on Social and Behavioral Computing
SMS	Short Message Service
SRI	Stanford Research Institute
TAM	Technology Acceptance Model
TED (talks)	Technology, Education, and Design
THCI	(AIS) Transactions on Human Computer Interaction
TOCHI	(ACM) Transactions on Computer-Human Interaction
TOOIS, TOIS	Transactions on Office Information Systems (dropped "Office" in 1991)
TX-0, TX-2	Transistorized eXperimental computers 0 and 2
UCLA	University of California, Los Angeles
UDC	Universal Decimal Classification (system)
UNIX	Uniplexed Information and Computing Service

UPA, UXPA	Usability Professionals Association (User Experience since 2012)
UX	User eXperience
VDT, VDU	Video Display Technology, Video Display Unit
VLSI	Very Large Scale Integration
(The) WELL	Whole Earth 'Lectronic Link
WYSIWYG	What You See Is What You Get
Y2K	Year 2000

Bibliography

Note: All URLs were accessed November 16, 2016. Any broken links can likely be found on the Wayback Machine, https://web.archive.org.

Ackoff, R. L. (1967). Management misinformation systems. *Management Science*, 14, B147–B156. DOI: 10.1287/mnsc.14.4.B147. 27

Allen, P. G. and Greaves, M. (2011). The Singularity isn't near. *MIT Technology Review*, October 12. https://www.technologyreview.com/s/425733/paul-allen-the-singularity-isnt-near/. 102

Asimov, I. (1950). *I, Robot*. New York: Gnome Press.

Aspray, W. (1999). Command and control, documentation, and library science: The origins of information science at the University of Pittsburgh. *IEEE Annals of the History of Computing*, 21(4), 4–20. DOI: 10.1109/85.801528. 23, 35

Baecker, R. and Buxton, W. (1987). A historical and intellectual perspective. In Baecker and Buxton, 1987, pp. 41–54.

Baecker, R. and Buxton, W. (Eds.) (1987). *Readings in HCI: A Multidisciplinary Approach*. San Francisco: Morgan Kaufmann. 27

Baecker, R. (Ed.). (1993). *Readings in Groupware and Computer-Supported Cooperative Work: Assisting Human-Human Collaboration*. Morgan Kaufmann.

Baecker, R., Grudin, J., Buxton, W. and Greenberg, S. (1995). A historical and intellectual perspective. In R. Baecker, J. Grudin, W. Buxton and S. Greenberg, *Readings in HCI: Toward the Year 2000* (pp. 35–47). San Francisco: Morgan Kaufmann.

Bagozzi, R. P., Davis, F. D. and Warshaw, P. R. (1992). Development and test of a theory of technological learning and usage. *Human Relations*, 45(7), 660–686. DOI: 10.1177/001872679204500702. 60

Banker, R. D. and Kaufmann, R. J. (2004). The evolution of research on Information Systems: A fiftieth-year survey of the literature in Management Science. *Management Science*, 50(3), 281–298. DOI: 10.1287/mnsc.1040.0206. 3, 27

Bannon, L. (1991). From human factors to human actors: The role of psychology and HCI studies in system design. In J. Greenbaum and M. Kyng (Eds.), *Design at Work* (pp. 25–44). Hillsdale, NJ: Erlbaum. 43, 48

Bardini, T. (2000). *Bootstrapping: Douglas Engelbart, Coevolution, and the Origins of Personal Computing*. Stanford University. 23

Barnard, P. (1991). Bridging between basic theories and the artifacts of HCI. In J. M. Carroll (Ed.), *Designing Interaction: Psychology at the Human-Computer Interface* (pp. 103–127). Cambridge: Cambridge University Press. 56

Barnard, P., May, J, Duke, D. and Duce, D. (2000). Systems, interactions, and macrotheory. *ACM Trans. Computer-Human Interaction*, 7(2), 222–262. DOI: 10.1145/353485.353490. 77

Begeman, M., Cook, P., Ellis, C., Graf, M., Rein, G., and Smith, T. (1986). Project Nick: Meetings augmentation and analysis. *Proc. Computer-Supported Cooperative Work 1986*, 1–6. DOI: 10.1145/637069.637071. 60

Benbasat, I. and Dexter A. S. (1985). An experimental evaluation of graphical and color-enhanced information presentation. *Management Science*, 31(11), 1348–1364. DOI: 10.1287/mnsc.31.11.1348. 59, 91

Bennett, J. L. (1979). The commercial impact of usability in interactive systems. In B. Shackel (Ed.), *Man-Computer Communication* (Vol. 2, pp. 1-17). Maidenhead: Pergamon-Infotech. 42

Bewley, W. L., Roberts, T. L., Schroit, D. and Verplank, W. L. (1983). Human factors testing in the design of Xerox's 8010 "Star" office workstation. *Proc. CHI'83*, 72–77. New York: ACM. DOI: 10.1145/800045.801584. 78

Beyer, H. and Holtzblatt, K. (1998). *Contextual Design—Defining Customer-Centered Systems*. San Francisco: Morgan Kaufmann. 93

Bjerknes, G., Ehn, P., and Kyng, M. (Eds.). (1987). *Computers and Democracy—a Scandinavian Challenge*. Aldershot, UK: Avebury. 64

Björn-Andersen, N. and Hedberg, B. (1977). Design of information systems in an organizational perspective. In P.C. Nystrom and W.H. Starbuck (Eds.), *Prescriptive Models of Organizations* (pp. 125–142). *TIMS Studies in the Management Sciences*, Vol. 5. Amsterdam: North-Holland. 27

Blackwell, A. (2006). The reification of metaphor as a design tool. *ACM Trans. Computer-Human Interaction*, 13(4), 490–530. DOI: 10.1145/1188816.1188820. 3, 123

Blythe, M. A., Monk, A. F., Overbeeke, K. and Wright, P. C. (Eds.). (2003). *Funology: From Usability to User Enjoyment*. New York: Kluwer. 79

Borman, L. (1996). SIGCHI: the early years. *SIGCHI Bulletin*, 28(1), 1–33. New York: ACM. DOI: 10.1145/249170.249172. 47

Bowman, W. J. (1968). *Graphic Communication*. New York: John Wiley. 78

Brynjolfsson, E. (1993). The productivity paradox of information technology. *Commun. ACM*, 36, 12, 66–77. DOI: 10.1145/163298.163309. 62

Brynjolfsson, E. and Hitt, L. M. (1998). Beyond the productivity paradox. *Commun. ACM*, 41, 8, 49–55. DOI: 10.1145/280324.280332. 62

Buckland, M. (1998). Documentation, information science, and library science in the U.S.A. In Hahn and Buckland, pp. 159–172. 9

Buckland, M. (2009). As we may recall: Four forgotten pioneers. *ACM Interactions*, 16(6), 76–69. DOI: 10.1145/1620693.1620712. 10

Burke, C. (1994). *Information and Secrecy: Vannevar Bush, Ultra, and the Other Memex*. Lanham, MD: Scarecrow Press. 3, 23

Burke, C. (1998). A rough road to the information highway: Project INTREX. In Hahn and Buckland, pp. 132–146. 35

Burke, C. (2007). History of information science. In Cronin, B. (Ed.), *Annual Review of Information Science and Technology 41*, pp. 3–53. Medford, NJ: Information Today/ASIST. DOI: 10.1002/aris.2007.1440410108. 3, 9

Bush, V. (1945). As we may think. *The Atlantic Monthly*, 176, 101–108. http://www.theatlantic.com/magazine/archive/1969/12/as-we-may-think/3881/. 10

Butler, P. (1933). *Introduction to Library Science*. Chicago: Univ. of Chicago Press. 9

Buxton, W. A. S. (2006). *Early Interactive Graphics at MIT Lincoln Labs*. http://www.billbuxton.com/Lincoln.html. 21, 57

Bødker, S. (1990). *Through the Interface: A Human Activity Approach to User Interface Design*. Mahwah, NJ: Lawrence Erlbaum. 120

Cakir, A., Hart, D. J., and Stewart, T. F. M. (1980). *Visual Display Terminals*. New York: Wiley. 26

Card, S. K. and Moran, T. P. (1986). User technology: From pointing to pondering. *Proc. Conference on the History of Personal Workstations*, 183–198. New York: ACM. DOI: 10.1145/12178.12189. 28

Card, S. K., Moran, T. P., and Newell, A. (1980a). Computer text-editing: An information-processing analysis of a routine cognitive skill. *Cognitive Psychology*, 12, 396–410. DOI: 10.1016/0010-0285(80)90003-1. 49

Card, S. K., Moran, T. P., and Newell, A. (1980b). Keystroke-level model for user performance time with interactive systems. *Comm. ACM*, 23(7), 396–410. New York: ACM. DOI: 10.1145/358886.358895. 49, 50

Card, S., Moran, T. P. and Newell, A. (1983). *The Psychology of Human-Computer Interaction*. Mahwah, NJ: Lawrence Erlbaum. 77

Carey, J. (1988). *Human Factors in Management Information Systems*. Greenwich, CT: Ablex. 60

Carroll, J. M. (Ed.) (2003). *HCI Models, Theories and Frameworks: Toward a Multidisciplinary Science*. San Francisco: Morgan Kaufmann. 77

Carroll, J. M. and Campbell, R. L. (1986). *Softening up Hard Science: Response to Newell and Card. Human-Computer Interaction*, 2(3), 227–249. DOI: 10.1207/s15327051hci0203_3. 56

Carroll, J.M. and Mack, R.L. (1984). Learning to use a word processor: by doing, by thinking, and by knowing. In J. Thomas and M. Schneider (Eds.), *Human Factors in Computer Systems*, pp. 13–51. Ablex. 2

Carroll, J. M. and Mazur, S. A. (1986). Lisa learning. *IEEE Computer*, 19(11), 35–49. DOI: 10.1109/MC.1986.1663098. 56

Christensen, C. (1997). T*he Innovator's Dilemma: When New Technologies Cause Great Firms to Fail*. Harvard Business Review Press. 38, 99

Cronin, B. (1995). Shibboleth and substance in North American Library and Information Science education. *Libri*, 45, 45–63. DOI: 10.1515/libr.1995.45.1.45. 67, 75, 86

Daft, R. L. and Lengel, R. H. (1986). Organizational information requirements, Media Richness and structural design. *Management Science*, 32, 5, 554–571. DOI: 10.1287/mnsc.32.5.554. 63

Damodaran, L., Simpson, A., and Wilson, P. (1980). *Designing Systems for People*. Manchester, UK: NCC Publications. 26

Darrach, B. (1970). Meet Shaky: The first electronic person. *Life Magazine*, 69(21), 58B–68. 32, 33

Davis, F. D. (1989). Perceived usefulness, perceived ease of use, and user acceptance of information technology. *MIS Quarterly*, 13(3), 319–339. DOI: 10.2307/249008. 60

Davis, G. B. (1974). *Management Information Systems: Conceptual Foundations, Structure, and Development*. New York: McGraw-Hill. 27

Dennis, A. R. and Reinicke, B. A. (2004). Beta versus VHS and the acceptance of electronic brainstorming technology. *MIS Quarterly*, 28(1), 1–20. 60

Dennis, A., George, J., Jessup, L., Nunamaker, J., and Vogel, D. (1988). Information technology to support electronic meetings. *MIS Quarterly*, 12(4), 591–624. DOI: 10.2307/249135. 60

DeSanctis, G. and Gallupe, R. B. (1987). A foundation for the study of group decision support systems. *Management Science*, 33, 589–610. DOI: 10.1287/mnsc.33.5.589. 60

Dourish, P., Adler, A., Bellotti, V., and Henderson, A. (1996). Your place or mine? Learning from long-term use of audio-video communication. *Computer-Supported Cooperative Work*, 5, 1, 33–62. DOI: 10.1007/BF00141935. 107

Dyson, F. (1979). *Disturbing the Universe*. New York: Harper and Row. 7

Ehrlich, S. F. (1987). Strategies for encouraging successful adoption of office communication systems. *ACM Trans. Office Information Systems*, 5(4), 340–357. DOI: 10.1145/42196.42198.

Ellickson, R. C. (1991). *Order Without Law*. Cambridge, MA:Harvard University Press. 108

Engelbart, D. (1962). Augmenting human intellect: A conceptual framework. SRI Summary report AFOSR-3223. Reprinted in P. Howerton and D. Weeks (Eds.), *Vistas in Information Handling*, Vol. 1 (pp. 1–29). Washington, D.C.: Spartan Books, 1963. http://www.dougEngelbart.org/pubs/augment-3906.html. 22

Engelien, B. and McBryde, R. (1991). *Natural Language Markets: Commercial Strategies*. London: Ovum Ltd. 52

Evenson, S. (2005). Design and HCI highlights. Presented at the *HCIC 2005 Conference*. Winter Park, Colorado, February 6, 2005. 78

Fano, R. and Corbato, F. (1966). Timesharing on computers. *Scientific American* 214(9), 129–140. 21

Farooq, U. and Grudin, J. (2016). Human-computer interaction integration. *ACM Interactions*, 23, 5, 26–32. DOI: 10.1145/3001896. 100

Feigenbaum, E. A. and McCorduck, P. (1983). *The Fifth Generation: Artificial Intelligence and Japan's Computer Challenge to the World*. Reading, MA: Addison-Wesley.

Fidel, R. (2011). *Human Information Interaction: An Ecological Approach to Information Behavior*. Cambridge, MA: MIT Press. 35

Fischer, G. and Nakakoji, K. (1992). Beyond the macho approach of Artificial Intelligence: Empower human designers - do not replace them. *Knowledge-Based Systems Special Issue on AI in Design*, 5, 1, 15–30. DOI: 10.1016/0950-7051(92)90021-7. 101

Foley, J. D. and Wallace, V. L. (1974). The art of natural graphic man-machine conversation. *Proc. of the IEEE*, 62(4), 462–471. DOI: 10.1109/PROC.1974.9450. 30

Forbus, K. (2003). Sketching for knowledge capture. Lecture at Microsoft Research, Redmond, WA, May 2. 80

Forbus, K. D., Usher, J., and Chapman, V. (2003). Qualitative spatial reasoning about sketch maps. *Proc. Innovative Applications of AI*, pp. 85-92. Menlo Park: AAAI. 80

Forster, E. M. (1909). The machine stops. *Oxford and Cambridge Review*, 8, November, 83–122. 36

Friedman, A. (1989). *Computer Systems Development: History, Organization and Implementation*. New York: Wiley.

Gilbreth, L. (1914). *The Psychology of Management: The Function of the Mind in Determining Teaching and Installing Methods of Least Waste*. NY: Sturgis and Walton.

Giles, J. (2005). Internet encyclopedias go head to head. *Nature*, 438, 7070, 900–901. DOI: 10.1038/438900a. 81

Glushko, R. J. (Ed.) (2016). *The Discipline of Organizing*, Professional edition (4th edition). O'Reilly Media. http://disciplineoforganizing.org/. 88, 112

Good, I. J. (1965). Speculations concerning the first ultra-intelligent machine. *Advances in Computers*, 6, 31–88. http://tinyurl.com/ooonkdu. DOI: 10.1016/S0065-2458(08)60418-0. 34

Gould, J. D. and Lewis, C. (1983). Designing for usability—Key principles and what designers think. Proc. *CHI'83*, 50–53. NY: ACM. DOI: 10.1145/800045.801579.

Grandjean, E. and Vigliani, A. (1980). *Ergonomics Aspects of Visual Display Terminals*. London: Taylor and Francis. 26

Gray, W. D., John, B. E., Stuart, R., Lawrence, D., and Atwood, M. E. (1990). GOMS meets the phone company: Analytic modeling applied to real-world problems. *Proc. Interact'90*, 29–34. Amsterdam: North Holland. 50

Greenbaum, J. (1979). *In the Name of Efficiency*. Philadelphia: Temple University. 26

Greif, I. (1985). Computer-Supported Cooperative Groups: What are the issues? *Proc. AFIPS Office Automation Conference*, 73-76. Montvale, NJ: AFIPS Press. 57, 62

Greif, I. (ed.) (1988). *Computer-Supported Cooperative Work: A Book of Readings*. San Mateo, CA: Morgan Kaufmann. 62

Grudin, J. (1990a). Groupware and cooperative work: Problems and prospects. In B. Laurel (Ed.), *The Art of Human-Computer Interface Design*, 171–185. Addison Wesley. Reprinted in Baecker (1995). 122

Grudin, J. (1990b). The computer reaches out: The historical continuity of interface design. *Proc. CHI'90*, 261–268. NY: ACM. DOI: 10.1145/97243.97284. 65

Grudin, J. (1991). Interactive systems: Bridging the gaps between developers and users. *IEEE Computer*, 24(4), 59–69. DOI: 10.1109/2.76263. 119

Grudin, J. (1993). Interface: An evolving concept. *Comm. ACM,* 36(4), 110–119. DOI: 10.1145/255950.153585. 120

Grudin, J. (2007). NordiCHI 2006: Learning from a regional conference. *ACM Interactions*, 14, 3, 52–53. DOI: 10.1145/1242421.1242458. 121

Grudin, J. (2009). AI and HCI: Two fields divided by a common focus. *AI Magazine*, 30(4), 48–57. 35

Grudin, J. (2011a). Human-computer interaction. In Cronin, B. (Ed.), *Annual Review of Information Dcience and Technology 45*, pp. 369–430. Medford, NJ: Information Today (for ASIST).

Grudin, J. (2011b). Technology, conferences, and community. *AI MagazineCommunications of the ACM*, 54, 2, 41–43. DOI: 10.1145/1897816.1897834. 93

Grudin, J. (2012). A moving target: The evolution of human-computer interaction. In *Jacko* (2012), pp. xvii–lxi).

Grudin, J. and MacLean, A. (1984). Adapting a psychophysical method to measure performance and preference tradeoffs in human-computer interaction. *Proc. INTERACT'84*, 338–342. Amsterdam: North Holland. 118

Hahn, T. B. and Buckland, M. (Eds.) (1998). *Historical Studies in Information Science*. Information Today/ASIS.

Helander, M. (Ed.) (1988). *Handbook of Human-Computer Interaction*. Amsterdam: North-Holland. 57

Helander, M. G., Landauer, T. K., and Prabhu, P. V. (Eds.) (1991) *Handbook of Human-Computer Interaction*, 2nd edition. Helander North-Holland.

Hertzfeld, A. (2005). R*evolution in the Valley: The Insanely Great Story of How the Mac Was Made*. Sebastopol, CA: O'Reilly Media.

HFES (2010). HFES history. In *HFES 2010–2011, Directory and Yearbook* (pp. 1–3). Santa Monica: Human Factors and Ergonomics Society. Also found at http://www.HFES.org/web/AboutHFES/history.html. 8

Hiltz, S. R. and Turoff, M. (1978). *The Network Nation*. Reading, MA: Addison-Wesley. 36

Hiltzik, M. A. (1999). *Dealers of Lightning: Xerox PARC and the Dawn of the Computer Age*. New York: HarperCollins. 30

Hopper, G. (1952). The education of a computer. Proc. *ACM Conference*, reprinted in Annals of the History of Computing, 9(3–4), 271–281, 1987. DOI: 10.1145/609784.609818. 16

Huber, G. (1983). Cognitive style as a basis for MIS and DSS designs: Much ado about nothing? *Management Science*, 29(5), 567–579. DOI: 10.1287/mnsc.29.5.567. 59

Hutchins, E. L., Hollan, J. D., and Norman, D. A. (1986). Direct manipulation interfaces. In D. A. Norman and S. W. Draper (Eds.), *User Centered System Design* (pp. 87–124). Mahwah, NJ: Lawrence Erlbaum. 56

Isaacson, W. (2014). *How a Group of Hackers, Geniuses, and Geeks Created the Digital Revolution*. Simon and Schuster. 58

Israelski, E. and Lund, A. M. (2003). The evolution of HCI during the telecommunications revolution. In Jacko and Sears (2003), pp. 772–789. 42

Jacko, J. (Ed.) (2012). *The Human-Computer Interaction Handbook: Fundamentals, Evolving Technologies, and Emerging Applications* (3rd edition). CRC Press. DOI: 10.1201/b11963.

Jacko, J.A. and Sears, A. (Eds.) (2003). *he Human-Computer Interaction Handbook: Fundamentals, Evolving Technologies, and Emerging Applications*. Lawrence Erlbaum.

Johnson, T. (1985). *Natural Language Computing: The Commercial Applications*. London: Ovum Ltd. 52

Kao, E. (1998). *The History of AI*. https://web.archive.org/web/20130101000000/http://www.generation5.org/content/1999/aihistory.asp. 52

Kay, A. (1969). The reactive engine. Ph.D. thesis, University of Utah. http://www.mprove.de/diplom/gui/kay69.html. 29

Kay, A. and Goldberg, A. (1977). Personal dynamic media. *IEEE Computer* 10(3), 31–42. DOI: 10.1109/C-M.1977.217672. 29

Keen, P. G. W. (1980). MIS research: reference disciplines and a cumulative tradition. In *First International Conference on Information Systems*, 9–18. Chicago: Society for Management Information Systems. 27

Kling, R. (1980). Social analyses of computing: Theoretical perspectives in recent empirical research. *Computing Surveys*, 12(1), 61–110. DOI: 10.1145/356802.356806. 27

Kowack, G. (2008). Unanticipated and contingent influences on the evolution of the internet. *ACM Interactions*, 15, 1, 74–78. DOI: 10.1145/1330526.1330551. 70

Kraemer, K. L. and King, J. L. (1988). Computer-based systems for cooperative work and group decision making. *ACM Computing Surveys*, 20, 2, 115–146. DOI: 0.1145/46157.46158. 60

Kraut, R., Kiesler, S., Boneva, B., Cummings, J. N., Helgeson, V., and Crawford, A. M. (2002). Internet paradox revisited. *Journal of Social Issues*, 58, 1, 49–74. DOI: 10.1111/1540-4560.00248. 62

Kraut, R., Patterson, M., Lundmark, V., Kiesler, S., Mukhopadhyay, T., and Scherlis, W. (1998). Internet paradox: A social technology that reduces social involvement and psychological well-being? *American Psychologist*, 53, 9, 1017–1031. DOI: 10.1037/0003-066X.53.9.1017. 62

Landau, R., Bair, J., and Siegmna, J. (Eds.). (1982). Emerging office systems: *Extended Proceedings of the 1980 Stanford International Symposium on Office Automation*, Norwood, NJ. 44

Lenat, D. (1989). When will machines learn? *Machine Learning*, 4, 255–257. DOI: 10.1007/BF00130713.

Levy, S. (1984). *Hackers: Heroes of the Computer Revolution*. Anchor Press/Doubleday. 47

Levy, S. (1994). *Insanely Great: The Life and Times of Macintosh, the Computer that Changed Everything*. Viking.

Lewis, C. (1983). The 'thinking aloud' method in interface evaluation. Tutorial given at *CHI'83*. Unpublished notes. 93

Lewis, C. and Mack, R. (1982). Learning to use a text processing system: Evidence from "thinking aloud" protocols. *Proc. Conference on Human Factors in Computing Systems*, 387–392. New York: ACM. DOI: 10.1145/800049.801817. 93

Lewis, C. Polson, P., Wharton, C., and Rieman, J. (1990). Testing a walkthrough methodology for theory-based design of walk-up-and-use Interfaces. *Proc. CHI'90*, 235–242. New York: ACM. DOI: 10.1145/97243.97279. 93

Licklider, J. C. R. (1960). Man-computer symbiosis. *IRE Transactions of Human Factors in Electronics HFE-1*, 1, 4–11. http://groups.csail.mit.edu/medg/people/psz/Licklider.html. DOI: 10.1109/THFE2.1960.4503259. 20, 99

Licklider, J.C.R. (1963). Memorandum for: Members and Affiliates of the Intergalactic Computer Network. ARPA, April 23. http://www.kurzweilai.net/memorandum-for-members-and-affiliates-of-the-intergalactic-computer-network.

Licklider, J. C. R. (1965). *Libraries of the Future*. Cambridge, MA: MIT Press.

Licklider, J. C. R. (1976). User-oriented interactive computer graphics. In *Proc. SIGGRAPH Workshop on User-Oriented Design of Interactive Graphics Systems*, 89–96. New York: ACM. DOI: 10.1145/1024273.1024284. 30

Licklider, J. C. R. and Clark, W. (1962). On-line man-computer communication. *AFIPS Conference Proc.*, 21, 113–128. DOI: 10.1145/1460833.1460847. 21

Lighthill, J. (1973). Artificial intelligence: A general survey. In J. Lighthill, N. S. Sutherland, R. M. Needham, H. C. Longuet-Higgins and D. Michie (Eds.), *Artificial Intelligence: A Paper Symposium*. London: Science Research Council of Great Britain. http://www.chilton-computing.org.uk/inf/literature/reports/lighthill_report/p001.htm.

Long, J. (1989). Cognitive ergonomics and human-computer interaction. In J. Long and A. White-field (Eds.), *Cognitive Ergonomics and Human-Computer Interaction* (pp. 4–34). Cambridge: Cambridge University Press. 56

Machlup, F. and Mansfield, U. (Eds.) (1983). *The Study of Information: Interdisciplinary Messages*. New York: Wiley.

March, A. (1994). Usability: the new dimension of product design. *Harvard Business Review*, 72(5), 144–149. 61

Marcus, A. (2004). Branding 101. *ACM Interactions*, 11(5), 14–21. DOI: 10.1145/1015530.1015539. 79

Markoff, J. (2005). *What the Dormouse Said: How the 60s Counter-Culture Shaped the Personal Computer*. London: Viking. 39

Markoff, J. (2015). *Machines of Loving Grace: The Quest for Common Ground between Humans and Robots*. New York: HarperCollins. 20, 35, 100

Markus, M. L. (1983). Power, politics, and MIS implementation. *Comm. of ACM*, 26(6), 430–444. DOI: 10.1145/358141.358148. 27

Martin, J. (1973). *Design of Man-Computer Dialogues*. New York: Prentice-Hall. 26

McCarthy, J. (1960). Functions of symbolic expressions and their computation by machine, part 1. *Comm. ACM*, 3(4), 184–195. DOI: 10.1145/367177.367199. 31

McCarthy, J. (1988). B. P. Bloomfield, The question of artificial intelligence: Philosophical and sociological perspectives. *Annals of the History of Computing*, 10(3), 224–229. http://www-formal.Stanford.edu/jmc/reviews/bloomfield/bloomfield.html. 31

McKeown, K. (1991). Social norms and implications of Santa Monica's PEN (Public Electronic Network). American Psychological Association presentation. http://www.mckeown.net/PENaddress.html. 109

Meister, D. (1999). *The History of Human Factors and Ergonomics*. Mahwah, NJ: Lawrence Erlbaum. 8, 76

Mintzberg, H. (1984). A typology of organizational structure. In D. Miller and P. H. Friesen (Eds.), *Organizations: A Quantum View* (pp. 68–86). Englewood Cliffs, NJ: Prentice-Hall. Reprinted in R. Baecker, 1993. 119

Moggridge, B. (2007). *Designing Interactions*. Cambridge: MIT Press.

Moody, F. (1995). *I Sing the Body Electronic: A Year with Microsoft on the Multimedia Frontier*. Viking. 78

Moravec, H. (1988). *Mind Children: The Future of Robot and Human Intelligence*. Cambridge: Harvard University Press. 34

Moravec, H. (1998). When will computer hardware match the human brain? *Journal Evolution and Technology*, 1, 1. http://www.transhumanist.com/volume1/moravec.htm. 32

Mumford, E. (1971). A comprehensive method for handling the human problems of computer introduction. *IFIP Congress*, 2, 918–923. 27

Mumford, E. (1976). Toward the democratic design of work systems. *Personnel Management*, 8(9), 32–35. 27

Myers, B. A. (1998). A brief history of human computer interaction technology. *ACM Interactions*, 5(2), 44–54. DOI: 10.1145/274430.274436. 3

National Science Foundation. (1993). *NSF 93–2: Interactive Systems Program Description*. 13 January 1993. http://www.nsf.gov/pubs/stis1993/nsf932/nsf932.txt. 59, 77

National Science Foundation. (2003). NSF Committee of Visitors Report: Information and Intelligent Systems Division. 28 July 2003.

Negroponte, N. (1970). *The Architecture Machine: Towards a More Humane Environment*. Cambridge, MA: MIT Press. 32

Nelson, T. (1965). Complex file structure: A file structure for the complex, the changing, and the indeterminate. *Proc. ACM National Conference*, 84–100. New York: ACM. DOI: 10.1145/800197.806036. 23

Nelson, T. (1973). A conceptual framework for man-machine everything. *Proc. National Computer Conference* (pp. M21–M26). Montvale, New Jersey: AFIPS Press. DOI: 10.1145/1499586.1499776. 23

Nelson, T. (1996). Generalized links, micropayment and transcopyright. http://tinyurl.com/jp-sq8bz. 23

Newell, A., Arnott, J., Dye, R., and Cairns, A. (1991). A full-speed listening typewriter simulation. *Intnl J Man-Machine Studies*, 35, 2, 119–131. << Note: This is a different A. Newell than the others. >> DOI: 10.1016/S0020-7373(05)80144-0. 92

Newell, A. and Card, S. K. (1985). The prospects for psychological science in human-computer interaction. *Human-Computer Interaction*, 1(3), 209–242. DOI: 10.1207/s15327051hci0103_1. 49, 50, 56

Newell, A. and Simon, H. A. (1956). The logic theory machine: A complex information processing system. *IRE Transactions on Information Theory* IT-2, 61–79. DOI: 10.1109/TIT.1956.1056797.

Newell, A. and Simon, H. A. (1972). *Human Problem Solving*. New York: Prentice-Hall. 93

Newman, W. M. and Sproull, R. F. (1973).*Principles of Interactive Computer Graphics*. New York: McGraw-Hill. 29

Nielsen, J. (1989). Usability engineering at a discount. In G. Salvendy and M.J. Smith (Eds.), *Designing and Using Human-Computer Interfaces and Knowledge Based Systems* (pp. 394–401). Amsterdam: Elsevier. 56, 93

Nielsen, J. and Molich, R. (1990). Heuristic evaluation of user interfaces. *Proc. CHI'90*, 249-256. New York: ACM. DOI: 10.1145/97243.97281. 56

Norberg, A. L. and O'Neill, J. E. (1996). *Transforming Computer Technology: Information Processing for the Pentagon* 1962–1986. Baltimore: Johns Hopkins. 28, 52

Norman, D. A. (1982). Steps toward a cognitive engineering: Design rules based on analyses of human error. *Proc. Conference on Human Factors in Computing Systems*, 378–382. New York: ACM. DOI: 10.1145/800049.801815. 77

Norman, D. A. (1983). Design principles for human-computer interfaces. *Proc. CHI'83*, 1–10. New York: ACM. DOI: 10.1145/800045.801571.

Norman, D. A. (1986). Cognitive engineering. In D. A. Norman and S. W. Draper (Eds.), *User Centered System Design* (pp. 31–61). Mahwah, NJ: Lawrence Erlbaum. 77

Norman, D. A. (1988). *Psychology of Everyday Things*. Reissued in 1990 as *Design of Everyday Things*. New York: Basic Books. 57, 85, 116

Norman, D. A. (2004). *Emotional Design: Why We Love (or Hate) Everyday Things*. New York: Basic Books. 85

Norman, D.A. and Draper, S. (Eds.) (1985). *User Centered System Design: New Perspectives on Human-Computer Interaction*. Lawrence Erlbaum Associates. 57, 85

Nunamaker, J. (2004). Opening remarks at HICSS-38, January 2004. 94

Nunamaker, J., Briggs, R. O., Mittleman, D. D., Vogel, D. R. and Balthazard, P. A. (1997). Lessons from a dozen years of group support systems research: A discussion of lab and field findings. *Journal of Management Information Systems*, 13(3), 163–207. DOI: 10.1080/07421222.1996.11518138. 60

Nygaard, K. (1977). Trade union participation. Presentation at *CREST Conference on Management Information Systems*. Stafford, UK. 27

Oakley, B. W. (1990). Intelligent knowledge-based systems—AI in the U.K. In R. Kurzweil (Ed.), *The Age of Intelligent Machines* (pp. 346–349). Cambridge, MA: MIT Press. 52

Olson, G. M. and Olson, J. S. (2012). Collaboration technologies. In Jacko (2012), pp. 549-564. DOI: 10.1201/b11963-28. 63

Olson. G. M., Zimmerman, A., and Bos, N. (Eds.) (2008). *Scientific Collaboration on the Internet.* MIT Press. DOI: 10.7551/mitpress/9780262151207.001.0001. 71

Owczarek, S. (2011). What is a "UX Unicorn"? Quora, November 12. https://www.quora.com/What-is-a-UX-Unicorn. 83

Palen, L. (2015). Frontiers in crisis informatics. *CHI 2015* social impact award presentation. https://portal.klewel.com/watch/webcast/chi-2015-special-sigchi-social-impact-award/talk/1. 126

Palen, L. and Grudin, J. (2002). Discretionary adoption of group support software. In B.E. Munkvold, *Implementing Collaboration Technology in Industry*, (pp. 159-190). London: Springer-Verlag. 121

Perlman, G., Green, G. K. and Wogalter, M. S. (Eds.) (1995). *Human Factors Perspectives on Human-Computer Interaction.* Santa Monica: Human Factors and Ergonomics Society. 3

Pew, R. (2003). Evolution of HCI: From MEMEX to Bluetooth and beyond. In Jacko and Sears (2003), pp. 1–17. 3, 24, 26

Proc. Joint Conference on Easier and More Productive Use of Computer Systems, Part II: Human Interface and User Interface. (1981). New York: ACM. DOI: 10.1145/800276.810998.

Pruitt, J. and Adlin, T. (2006). *The Persona Lifecycle: Keeping People in Mind Throughout Product Design.* Morgan Kaufmann. 93

Rasmussen, J. (1980). The human as a system component. In H.T. Smith and T.R.G. Green (Eds.), *Human Interaction with Computers*, pp. 67-96. London: Academic.

Rasmussen, J. (1986). *Information Processing and Human-Machine Interaction: An Approach to Cognitive Engineering.* New York: North-Holland. 64

Rayward, W. B. (1983). Library and information sciences: Disciplinary differentiation, competition, and convergence. Edited section of F. Machlup and U. Mansfield (Eds.), *The Study of Information: Interdisciplinary Messages* (pp. 343–405). New York: Wiley. 3

Rayward. W. B. (1998). The history and historiography of Information Science: Some reflections. In Hahn and Buckland, pp. 7–21. 3

Remus, W. (1984). An empirical evaluation of the impact of graphical and tabular presentations on decision-making. *Management Science*, 30, 5, 533–542. DOI: 10.1287/mnsc.30.5.533. 59

Resnick, P., Iacovou, N., Suchak, M., Bergstrom, P. and Riedl, J. (1994). GroupLens: an open architecture for collaborative filtering of netnews. *Proc. CSCW 1994*, 175–186. DOI: 10.1145/192844.192905. 71

Roland, A. and Shiman, P. (2002). *Strategic Computing: DARPA and the Quest for Machine Intelligence*, 1983–1993. MIT Press. 59

Roscoe, S. N. (1997). *The Adolescence of Engineering Psychology*. Santa Monica, CA: Human Factors and Ergonomics Society. DOI: 10.1037/e721682011-001. 7, 8

Salvendy, G. (Ed.) (2012). *Handbook of Human Factors and Ergonomics* (4th edition). Wiley.

Sammet, J. (1992). Farewell to Grace Hopper—End of an era! *Comm. ACM,* 35(4), 128–131. DOI: 10.1145/129852.214846. 16

Schank, R. (2015). The fraudulent claims made by IBM about Watson and AI. Blog post. http://educationoutrage.blogspot.com/2015/11/the-fraudulent-claims-made-by-ibm-about.html. 102

Sears, A. and Jacko, J. A. (2007). *The Human-Computer Interaction Handbook: Fundamentals, Evolving Technologies. and Emerging Applications* (2nd edition). CRC Press. DOI: 10.1201/9781410615862.

Shackel, B. (1959). Ergonomics for a computer. *Design*, 120, 36–39. 20

Shackel, B. (1962). Ergonomics in the design of a large digital computer console. *Ergonomics*, 5, 229–241. DOI: 10.1080/00140136208930578. 20

Shackel, B. (1997). HCI: Whence and whither? *Journal of ASIS*, 48(11), 970–986. 3, 19, 20, 80

Shannon, C. E. (1950). Programming a computer for playing chess. *Philosophical Magazine*, 7(41), 256–275. DOI: 10.1080/14786445008521796.

Shannon, C. E. and Weaver, W. (1949). *The Mathematical Theory of Communication*. Urbana: Univ. of Illinois Press. 15

Sheil, B. A. (1981). The psychological study of programming. *ACM Computing Surveys*, 13(1), 101–120. DOI: 10.1145/356835.356840. 28

Shneiderman, B. (1980). *Software psychology: Human Factors in Computer and Information Systems*. Cambridge, MA: Winthrop. 28

Shneiderman, B. (1986). *Designing the User Interface: Strategies for Effective Human–Computer Interaction*. Addison-Wesley. 57

Simon, H. A. (1960). The new science of management decision. New York: Harper. Reprinted in *The Shape of Automation for Men and Management*, Harper and Row, 1965. DOI: 10.1037/13978-000. 34

Simon, H. A. (1981). *The Sciences of the Artificial*, second edition. MIT Press. 110

Smith, A. (2004). A history of computers at Georgia Tech. Unpublished manuscript. 24

Smith, D. K. and Alexander, R. C. (1999). *Fumbling the Future: How Xerox Invented, then Ignored, the First Personal Computer*. iUniverse.com.

Smith, H. T. and Green, T. R. G. (Eds.). (1980). *Human Interaction with Computers*. Orlando, FL: Academic. 42

Smith, S. L. (1963). Man-computer information transfer. In J. H. Howard (Ed.), *Electronic Information Display Systems* (pp. 284–299). Washington, DC: Spartan Books. 20

Smith, S. L., Farquhar, B. B. and Thomas, D. W. (1965). Color coding in formatted displays. *Journal of Applied Psychology*, 49, 393–398. DOI: 10.1037/h0022808. 25

Smith, S. L. and Goodwin, N. C. (1970). Computer-generated speech and man-computer interaction. *Human Factors*, 12, 215–223. 25

Smith, S. L. and Mosier, J. N. (1986). *Guidelines for Designing User Interface Software* (ESD-TR-86-278). Bedford, MA: MITRE. 58

Suchman, L. (1983). Office procedure as practical action: Models of work and system design. *ACM Trans. on Office Information Systems*, 1, 4, 320–328. DOI: 10.1145/357442.357445. 118

Suchman, L. (1987). *Plans and Situated Action: The Problem of Human-Machine Communication*. Cambridge University Press. 57

Suchman, L. (1988). Designing with the user: Review of 'Computers and democracy: A Scandinavian challenge. *ACM TOOIS*, 6, 2, 173–183. DOI: 10.1145/45941.383895. 64

Sutherland, I. (1963). Sketchpad: A man-machine graphical communication system. Doctoral dissertation, MIT. http://www.cl.cam.ac.uk/techreports/UCAM-CL-TR-574.pdf. 21

Taylor, F. W. (1911). *The Principles of Scientific Management*. New York: Harper.

Taylor, R. S. (1986). *Value-Added Processes in Information Systems*. Ablex. 67

Tesla Motors. (2016). *Full Self-Driving Hardware on All Cars*. https://www.tesla.com/autopilot. 102

Turing, A. (1949). Letter in London Times, June 11. See *Highlights from the Computer Museum report* Vol. 20, Summer/Fall 1987, p. 12. http://ed-thelen.org/comp-hist/TCMR-V20.pdf. 31, 32

Turing, A. (1950). Computing machinery and intelligence. Mind, 49, 433–460. Republished as "Can a machine think?" in J. R. Newman (Ed.), *The World of Mathematics*, Vol. 4 (pp. 2099–2123). New York: Simon and Schuster. DOI: 10.1093/mind/LIX.236.433. 31

Vessey, I. and Galletta, D. (1991). Cognitive fit: An empirical test of information acquisition. *Information Systems Research*, 2(1), 63–84. DOI: 10.1287/isre.2.1.63. 59

Waldrop, M. M. (2001). *The Dream Machine: J.C.R. Licklider and the Revolution that Made Computing Personal*. New York: Viking. 20

Weinberg, G. (1971). *The Psychology of Computer Programming*. New York: Van Nostrand Reinhold. 27

Weiser, M. (1991). The computer for the 21st century. *Scientific American*, 265, 3, 94–104. DOI: 10.1038/scientificamerican0991-94. 116

Wells, H. G. (1905). *A Modern Utopia*. London: Jonathan Cape. http://www.gutenberg.org/etext/6424.

Wells, H. G. (1938). *World Brain*. London: Methuen. 75

Wersig, G. (1992). Information science and theory: A weaver bird's perspective. In P. Vakkari and B. Cronin (Eds.), *Conceptions of Library and Information Science: Historical, Empirical, and Theoretical Perspectives* (pp. 201–217). London: Taylor Graham. 86

White, P. T. (1970). Behold the computer revolution. National Geographic, 38, 5, 593-633. http://blog.modernmechanix.com/2008/12/22/behold-the-computer-revolution/. 32

Wiggins, A. and Sawyer, S. (2012). Intellectual diversity in iSchools, evidence from faculty composition. *Journal of the American Society for Information Science and Technology*, 63(1), 8–21. DOI: 10.1002/asi.21619. 87

Williams, G. (1997). Task conflict and language differences: opportunities for videoconferencing? *Proc. ECSCW'97*, 97–108. DOI: 10.1007/978-94-015-7372-6_7. 107

Yates, J. (1989). *Control through Communication: The Rise of System in American Management*. Baltimore: Johns Hopkins. 8

Zhang, P. (2004). AIS SIGHCI three-year report. *SIGHCI Newsletter*, 3(1), 2–6. 74

Zhang, P., Nah, F. F.-H., and Preece, J. (2004). HCI studies in management information systems. *Behaviour and Information Technology*, 23(3), 147–151. DOI: 10.1080/01449290410001669905. 27, 74

Zhang, P., Li, N., Scialdone, M. J., and Carey, J. (2009). The intellectual advancement of Human-Computer Interaction Research: A critical assessment of the MIS Literature. *AIS Trans. Human-Computer Interaction*, 1, 3, 55–107. 3, 88

Author Biography

Jonathan Grudin's history connected to computer history on a school field trip, when he played black-jack against a huge computer at Battelle Laboratories in Columbus, Ohio. Perhaps a vacuum tube had burned out. The computer claimed to have won when it had not, convincing Jonathan that computers were impressive but not infallible. While in high school, he taught himself to write programs, enter them on punch cards, and run them on a nearby college's sole computer, unused in the evenings and on weekends. His first program found twin primes; his second constructed random bridge hands.

Jonathan majored in mathematics-physics at Reed College and obtained an M.S. in mathematics at Purdue University. After working as a programmer at Wang Laboratories and Stanford University, he obtained a Ph.D. in cognitive psychology, working with Don Norman at the University of California, San Diego and spending two of his summers at the MIT Artificial Intelligence Laboratory. During a two-year postdoc at the Medical Research Council Applied Psychology Unit in Cambridge, he conducted his first human-computer interaction studies with Phil Barnard and Allan MacLean. He returned to work as a software engineer at Wang Laboratories and team leader at the artificial intelligence-oriented Microelectronics and Computer Technology Corporation consortium in Austin, Texas. From 1989 to 1991, he was a visiting professor at Aarhus University. His first exploration of the field's history, in the library next to his office, was to determine the origin of the waterfall model of software development. From 1992 to 1998, he was a professor of information and computer science at the University of California, Irvine. He spent six-month sabbaticals at Keio University and the University of Oslo. In 1998, he joined Microsoft and became an affiliate professor at the University of Washington Information School.

Jonathan is an ACM Fellow and member of ACM SIGCHI's CHI Academy. He has participated in CHI and CSCW since their first conferences. His CSCW 1988 paper on the challenges in designing technology to support groups won the first CSCW Lasting Impact Award in 2014.

Index

Printed in the United States
by Baker & Taylor Publisher Services